A SPECIAL FORCES
AND CIA SOLDIER'S
FIFTY YEARS
IN THE FRONTLINES
OF THE WAR
AGAINST TERRORISM

HUNTING

THE

JACKAL

BILLY WAUGH
WITH
TIM KEOWN

AVON

U.S. $7.99
CAN. $10.99

ISBN 0-06-056410-5

9 780060 564100

5 0 7 9 9 >

EAN

HUNTING THE JACKAL

"Billy Waugh is the archetypal American Warrior Hero and *Hunting the Jackal* reads like a black ops textbook. It's a detailed, gripping inside look at the life-and-death shadow-wars against terrorism currently being fought in the mean streets and back alleys all over the world."
John Weisman, *New York Times* bestselling author of *SOAR* and *Jack in the Box*

"Tough, rambunctious, and smart, Sergeant Major Billy Waugh is a Special Forces cyclone whose career cut a powerful path in the world of covert operations. I'm honored to have served with him in SOG—and thankful he was on our side. His memoirs are essential reading to understand the real history of American Special Ops over the past century."
John L. Plaster, U.S. Army Special Forces, author of *SOG* and *Secret Commandos*

"For those of us who have served with Sgt. Maj. Billy Waugh it is gratifying to know that some of [his] incredible exploits have finally been recorded and made available to the public. This book, in addition to telling what missions were undertaken, provides a cold hard look at the reality our Special Operations men face every day. The stories here will give readers a real sense of what it is like to be in the trenches, or hiding in an improvised dugout surveilling, or going undercover in hostile foreign countries."
Major General John K. Singlaub, U.S. Army (Ret.), OSS, CIA, SOG

"*Hunting the Jackal* is a story of warriors told by the consummate warrior who has done it all. The exploits of Special Forces and their impact on warfare have only recently begun to be revealed to the general public. The odds, death, and destruction they face are beyond what the typical human psyche can deal with. Billy's stories are of valor and of warriors engaged in brutal warfare where their focus is always 'the Mission, the Mission, the Mission.' . . . The reader will quickly identify Waugh as a man with the 'right stuff' who had a rendezvous with destiny for which freedom and the United States are the benefactors."

Lew Merletti, former director, United States Secret Service, and former 5th Special Forces Group (Vietnam, 1969–1970)

"[Waugh is] a legend in the special operations world . . . His book traces a remarkable fifty-year career."

Washington Times

"Sgt. Maj. Billy Waugh is the man I want to cover my back in combat. His natural skills, ethics and loyalty have no equal, and he is an elite warrior who has made special operations a success over the past few decades. It is about time that his services to our country be recognized as he is the kind of man who should be a role model to the new generation of special operations soldiers."

Col. Sully H. de Fontaine, U.S. Army (Ret.), SOE, SAS, OSS, SF, SOG

HUNTING
THE
JACKAL

A SPECIAL FORCES AND CIA SOLDIER'S
FIFTY YEARS ON THE FRONTLINES
OF THE WAR AGAINST TERRORISM

BILLY WAUGH
WITH
TIM KEOWN

AVON BOOKS
An Imprint of HarperCollinsPublishers

AVON BOOKS
An Imprint of HarperCollins*Publishers*
10 East 53rd Street
New York, New York 10022-5299

Copyright © 2004 by Billy Waugh
All maps and sketches by Billy Waugh.
All photographs courtesy of Billy Waugh.
ISBN: 0-06-056410-5
www.avonbooks.com

First Avon Books paperback printing: June 2005
First William Morrow hardcover printing: July 2004

Avon Trademark Reg. U.S. Pat. Off. and in Other Countries,
Marca Registrada, Hecho en U.S.A.
HarperCollins® is a registered trademark of HarperCollins Publishers Inc.

Printed in the U.S.A.

10 9 8 7 6 5 4 3 2 1

To my wonderful Special Forces friends, living and dead.
You were my mentors.

You have never lived until you have almost died. For those who have fought for it, life has special flavor the protected will never know.

—SOA Creed

SGM Felipe Ahumada (Ret)

SGM Henry Bailey (Ret)

COL Aaron Bank (Deceased) "The Father of the Army Special Forces"

MG Eldon Bargewell (Active Duty)

COL Charles ("Charging Charlie") Beckwith (Deceased)

MSG Brooke Bell (Ret)

LTG Jerry Boykin (Active Duty)

SGM Vernon Broad (Ret)

GEN Bryan Brown (CO of SOCOM)

SGM Harry Brown (Deceased)

CPT Jim Butler (SOA #1)

MAJ Isaac Camacho (Ret) (Ex-POW)

MSG Arthur D. Childs (Deceased)

MSG Henry Corvera (Ret)

MSG Darren Crowder (Active Duty)

CSM Paul Darcy (Deceased)

COL Paris Davis (Ret)

MSG Jimmy Dean (Ret)

SGT Dale Dehnke (KIA)

SMA George Dunaway (Ret)

LTC Lee Dunlap (Ret)

MSG Wendell Enos (Ret)

COL/DR Warner D. ("Rocky") Farr (Active Duty)

SSG Donald Fawcett (Deceased)

MSG James ("Butch") Fernandez (Ret)

COL Sully de Fontaine (Ret)

SGM Alex Fontes (Deceased)

CWO4 Bill Fraiser (Ret)

CSM John Fryer (Ret)

SGM Wiley Gray (Deceased)

CSM Sammy Hernandez (Ret)

SFC Melvin Hill (Ret)

MAJ Jerry Kilburn (Deceased)

SFC Bruce Luttrell (KIA)

MSG Larry Manes (Ret)

COL O. Lee Mize (Ret) (Medal of Honor)

MSG Robert Moberg (Ret)

CSM Peter Morakon (Ret)

SFC David Morgan (KIA)

SFC Cliff Newman (Ret)

MSG Richard Norris (Ret)

COL Charles Norton (Ret)

MSG Richard Pegram (KIA)

COL Roger Pezzelle (Deceased)

MSG Angel Quisote (Ret)

MSG Marcus ("Pappy") Reed (Ret)

MG Ed Scholes (Ret)

COL Daniel Schungle (Deceased)

SGM Jimmy Scurry (Ret)

LTC William Shelton (Ret)

SGM Walter Shumate (Deceased)

MAJ Alan Shumate (Active Duty)

COL Arthur D. ("Bull") Simons (Deceased)

MAJ Clyde Sincere (Ret)

MG John Singlaub (Ret)

CSM Jack Smythe (Ret)

LTC Harlow Stevens (Ret)

MSG Howard Stevens (Ret) (Ex-POW)

SGT Madison Strohlein (KIA)

LTC Bill Sylvester (Ret)

LTG William Tangney (Ret)

MSG Paul Tracy
(Deceased)

LTC Larry Trapp
(Deceased)

SGM Art Tucker (Ret)

LTC James D.
("Shrimpboat")
Van Sickle
(Deceased)

MSG Charles Wesley
(Ret)

PO/1C (UDT) Clarence
Williams (Ret)

SGM Jack Williams
(Deceased)

CWO4 Ronald Wingo
(Ret)

MSG Jason T.
Woodworth (Ret)

SGM Fred Zabitosky
(Deceased) (Medal of
Honor)

And to the wonderful people of the CIA.
Keep up the fine work, boys and girls.

We sleep safe in our beds because rough men stand ready in the night to visit violence on those who would do us harm.

—GEORGE ORWELL

Then I heard the voice of the Lord saying, "Whom shall I send? And who will go for us?" And I said, "Here I am. Send me!"

—ISAIAH 6:8

PREFACE

On December 1, 2001, I celebrated my seventy-second birthday on the ground in Afghanistan, with a *chitrali* covering my head and an M-4 carbine slung over my right shoulder. I was beyond cold, and I stunk in a way civilized humans are not meant to stink. The contents of my nose flowed ceaselessly into a scraggly beard that was supposed to make me look more like a local Afghani and less like a freezing old man from Texas. The Vietnam-era shrapnel that resides in my knees and ankles felt like a bunch of frozen coins, and I was losing weight as if it were falling off my body. I was part of Team Romeo, a combined Special Forces/CIA takedown unit hunting Taliban and al Qaeda at nine thousand feet of elevation and -10°C, through the desolate high plains of southeastern Afghanistan. I was cold, dirty, and miserable, and I wouldn't have traded places with anybody in the world.

Two weeks earlier, when the United States Air Force C-17 headed for Afghanistan lifted off with me aboard, our country was officially embarking on its War on Terror. I, however, had been at war against terror for quite some time. To me, Operation Enduring Freedom was a natural extension of the work I'd been conducting for close to fifty years.

The men of Team Romeo, composed of CIA members and Special Forces ODA 594, treated me like a display in a war museum. They asked me to pose for photographs. They asked me about my experiences in Korea, my seven and a half years in Vietnam, my work on many high-profile operations as an independent contractor with the CIA. This attention made me uncomfortable and slightly embarrassed; I had convinced the higher-ups in my outfit that I could withstand the mental and physical punishment that awaited me in the first installment of the War on Terror, and I was not here to play the role of a living relic of the past half-century of U.S. combat. I was there to fight the Taliban and seek out al Qaeda, so when the adulation got too thick, I deflected it by saying, "Now, men, I assure you I cannot walk on water. And even if I could, there's none out here in the middle of this fucking desert."

Let me be clear: I am not a hero.

Chances are you have never heard my name, but I have worked in the shadows and along the margins of some of the most significant military and espionage events of the past fifty years. I have pursued enemies of the United States in sixty-four countries over those fifty years. I have faced danger in its many forms: armed, angry humans; sophisticated, undiscriminating weapons; harsh, unyielding landscapes.

There are many missions that cannot be recounted in the pages of a book, not ever. A good portion of my life has been classified, locked up in a safe in Langley or inside my memory. I will not betray any ongoing operations or threaten the lives of any of the great men who continue to fight those who have our demise as their ultimate goal. After all, I know the feeling. Through five decades of service to my country, I have purposefully and continuously placed myself in dangerous situations against our enemies. I have made my personal safety a secondary issue to the task at hand.

I have lived life on the edge of danger and of the law. I have found that I am good at it, and that I like it. I have developed qualities that are unique to my position; namely, I have lied my ass off many times to protect myself and my men. I have learned to avoid questions and suspicions from police or security forces in nations where I work.

Total countries in which I have worked: sixty-four.

Total times hauled in by unfriendly governments for spying: zero.

Total times tailed by unfriendlies, in their nation: countless.

Not each assignment was filled with excitement, and not every country tells a breathless story of dodging bullets and nabbing bad guys. But I have had my share of successes. I have been awarded one Silver Star, four Bronze Stars for Valor, four Commendation Ribbons for Valor, fourteen Air Medals for Valor, and two Combat Infantryman Badges. (Along the way I developed a propensity for attracting gunshots and shrapnel; I possess eight Purple Hearts to commemorate those occasions.) I joined the U.S. Army in 1947 and Special Forces in 1954, two years after its inception. Following my retirement from Special Forces, I embarked on a second career as a CIA independent contractor, hunting down some of the most notorious enemies of the United States.

I was one of the first CIA operatives to be assigned to keep tabs on Usama bin Laden in Khartoum, Sudan, in 1991 and 1992. In 1994, again in Khartoum, I was the leader of a four-man CIA team that conducted an epic search and surveillance operation that led to the capture of Carlos the Jackal. My job required me to inhabit the minds of these men and countless others, adopt their ways, see the world through their twisted eyes. I was forced to become a cultural chameleon, able to anticipate and understand the actions of men who are different from me in every aspect save one: dedication to a cause. Like any man who studies the tactics of his enemies and attempts to predict his actions, I gained a grudging respect for the men I hunted.

I attribute my accomplishments to hard work and persistence, old-fashioned concepts that never failed me. Perseverance is my best quality, to the point where it sometimes becomes a fault. The glimpses into my private life within these pages are few. My chosen lines of work—Special Forces, then CIA—are not conducive to long marriages or stable home lives. Whenever I was faced with a decision between home and work, there was very little debate. I chose work.

This dedication to my country and its protection took hold one week after my twelfth birthday, when I was interrupted from my job as a popcorn popper at the Strand Theater in Bastrop, Texas, by Bastrop County sheriff Ed Cartwright. It was a little after 2 p.m., a Sunday afternoon, and this larger-than-life man—the stereotypical hard-nosed Texas sheriff—strode into the theater lobby and said, "Billy, go upstairs and tell the projectionist to shut off the movie and turn on the house lights. And hurry up."

I did as I was told, leaving my dime-a-bag popcorn stand and heading up the stairs two at a time. The Movietone news for this day—December 7, 1941—had just ended and the projectionist was starting the prefeature comedy when I barged into the booth and repeated Sheriff Cartwright's instructions. The projectionist looked at me a little funny, but he followed orders as soon as he heard they came from the sheriff. Everyone in Bastrop County knew better than to fool with Ed Cartwright.

When the projectionist flipped on the house lights, I stepped onto the balcony area, where I looked down and watched the fifty or sixty patrons squint and grumble about the movie coming to a whirring halt. Sheriff Cartwright walked onto the stage and quieted the room by saying, "Folks, I have some news for you. You need to listen."

He was deadly serious, and every eye in the room centered on this tall, stern sheriff.

"Now," he continued, taking a deep breath. "The Japanese have bombed Pearl Harbor and done great damage to the United States Navy." He looked at the crowd. The importance of his words had yet to register. "Folks, it is being said the Japanese may bomb our country today or tomorrow. They may even invade with ships and men. I want you folks to know I am serious:

"We are at war as of this date."

That woke everybody up, even the teenagers who had been necking in the back. A man's voice broke the quiet by asking, "Mr. Ed, where's Pearl Harbor?"

"Hawaii," the sheriff said.

Another man yelled, "Well, Sheriff, where in the dang hell is Hawaii?"

I'm not sure Cartwright knew, at least not precisely, so he said, "Don't worry about Hawaii. Just walk from this movie house and go home. When you get there, place black coverings over your windows so no light shows."

As he walked off the stage he stopped and turned back. "And I don't want to see anyone on the streets. If you're not in your homes within one hour, I am going to put you in your house— forcefully."

I stayed behind the popcorn stand as the patrons filed out of the theater. There were no sales.

I went home and recounted the scene to my mother. She had not heard the news, but as an educated woman—a substitute teacher in our community—she knew the location of Pearl Harbor. Lillian Waugh dutifully cut up some black material and covered the windows of the small, one-bedroom apartment I shared with her and my older sister, Nancy.

This day is etched deeply and vividly in my mind. It wasn't fear I felt; it was excitement. Even at twelve years old, I itched to be part of the war. I would defend my country against its enemies, wherever and whoever they might be.

My father had died two years before, and I was consumed with the idea of duty. At least one man in each family had an obligation to perform for his country, and I was the only son of John and Lillian Waugh. Being a man in southwest Texas, to my way of thinking, meant being a military man.

I was ready for the military long before it was ready for me.

In 1945, just before my sixteenth birthday, another event in Bastrop shaped my future. Two local Marines, both wounded in World War II, returned to our little town. One of the Marines had a shrapnel wound to his head, the other wore a cast on his lower leg after receiving gunshot wounds in the Pacific. Whenever I was near them on the street or in a store, I felt awed to be in their presence. I admired their strength and nobility. They had seen things I could only dream of seeing, and I made a decision right then and there: I wanted to be like them. What they had done for the country, I would do for my country.

I knew enough to know I couldn't join the Marines in Texas, but I had heard somewhere that in Los Angeles a boy could join the Marines at fifteen or sixteen. I don't remember where I heard this information, but I do know I didn't question it. Instead, I ran away, hitchhiking west from Bastrop, with my destination some unknown recruiting station in a faraway, exotic city.

I made it as far as Las Cruces, New Mexico. Hitchhiking on the west side of town, I was approached by a local police officer.

"Where are you going, son?" the officer asked. "And what are you doing?"

I put on my best adult voice and said, "Going to Los Angeles, sir, to join the Marines."

I had no identification, and I refused to tell the cops where I came from. I was bold enough to attempt such a caper, but I knew full well my mother would have tanned my hide if she knew what I was doing.

"You look a little young for the Marine Corps," the officer told me.

Then, without giving me much chance to state my case, he threw me in jail.

There were some wild ones in the Las Cruces jail, and I quickly realized this wasn't the life for me. I pleaded with the police to let me go, that my only crime was a desire to serve my country. Eventually, I convinced the police to allow me to see a Marine recruiter in nearby Deming.

The recruiter looked at my skinny self and laughed in my face.

Then he asked me how old I was.

"I'm eighteen, sir," I said.

More laughter.

"Actually, I'll be eighteen soon," I said.

More laughter.

"Where's your mother, son?" the recruiter asked.

I gave him some vague answer, and it was back to jail for me. The police told me they wouldn't release me until I had a ticket out of town, and my complete lack of money made that impossible. The way my sixteen-year-old mind saw it, I was going to have to face my mother or spend the rest of my life with all these crazy men in the Las Cruces jail.

So I called my mother, explained the situation, and then listened as she gave me several of the sharper pieces of her mind. But by the end of the phone call she had agreed to wire me the bus fare from Las Cruces to Austin, and soon I was out of jail and on my way home. When I got there, my mother gave me a lengthy lecture and a firm belt-whipping. Also, a very clear set of orders: Get back in school, or else.

I finished high school with a 4.0—all A's—and a sore rear. Mom never did spare the belt on me, but she combined discipline with sound teachings on good manners, accepting responsibility, and understanding the importance of striving to achieve success in life.

The sting from the whippings eventually dissipated, but my desire to be part of the military did not. In August 1948, six months after my eighteenth birthday, I joined the U.S. Army paratroopers, a group I studied after hearing of their exploits during World War II. Beginning with my training at Fort Benning, I jumped out of a heck of a lot of aircraft. In fact, I didn't experience a landing until I returned from Korea in 1952—almost five years after I joined the military. I've jumped out of just about every aircraft possible—C-46s, C-47s, C-82s, C-119s, C-123s, C-130s, L-20s, Twin Beeches, U.S. Navy TF-1s, an Army Caribou, and helicopters by the dozen. I've crashed in three helicopters and two planes, but I've somehow managed to avoid death.

From Korea to Afghanistan and every conflict in between, I have fought whomever my country ordered me to fight. For fifty years in sixty-four countries, I have sought and destroyed my country's enemies—whether they be called Communists or terrorists—wherever they hide.

So this is my story—a Special Forces soldier's story—and the story of other soldiers. I tell their stories and mine as I remember them, and I hope I have done them justice. In many instances, names have been changed to protect the identities of those who remain involved in covert operations.

I have a story to tell, and I am finally ready to tell it.

CHAPTER 1

As I waited to die in a rice paddy in Bong Son, South Vietnam, on June 18, 1965, with green North Vietnamese Army (NVA) tracers searing past my naked, immobile body, my mind was not occupied by fear or regret. No, I drifted in and out of consciousness, my body perforated with gunshot wounds, leeches feasting on every open wound, with one thought jabbing at my semilucid brain: *Damn, my military career is finished. I'll never see combat again.*

Through eleven years in Special Forces and twenty-seven months in Southeast Asia, I had never been bashful when it came to combat. I lived for it, studied it, and understood it. I knew the risks and did not fear death. Still, I had never come close to being in a spot like this—flat on my back, shot to hell, lying behind a meager bamboo stand that provided pathetic protection. I was out of ammunition and gear. I had taken bullets to my knees, an arm, an ankle, a foot, and my forehead. The bones of my right foot and ankle sat there fully exposed, doing me absolutely no good while causing a breathtaking amount of pain. The force of one of the bullets had driven the sole of my right jungle boot through my foot and ankle and into my tibia. I could not crawl, let alone walk. The enemy had already gotten to me, stripping me and leaving me for dead. In this state, I apparently was not deemed worthy of the extra bullet that would have clinched my death. I was all alone, not a friendly in sight. There was no assurance that I would ever leave this bloody field or see the world from an upright position again.

And still the NVA kept firing. We had pissed the bastards off something fierce, and they weren't going to stop until every last one of us was as dead as I appeared to be. Their infernal green tracers were whizzing over my head, mocking my defenselessness, popping like cannon fire around my head as they broke the sound barrier. The kerosene smell and blast-furnace heat of the napalm blanketed that rice paddy, brought there by the Air Force F-4C Phantom and Navy F-8 jets screaming above.

When I took stock of my own dire predicament, peering through the now-crusted blood from the wound that had torn open my forehead, comprehending my utter nakedness, wondering how and why I continued to live, I began to ask myself a different question: When all this is over, how in the hell am I ever going to con my way back to the battlefield?

* * *

Getting into the battlefield was all that ever mattered to me. From the moment I joined the U.S. Army as an eighteen-year-old, I have never been content to sit back and hear of others' exploits. My desire to be among the troops at the point of attack struck me first in early 1951, when we were at war in Korea and I was stuck in the 82nd Airborne Division at Fort Bragg, North Carolina. I had had more than enough of the 82nd Abn. Div. and was tired of stateside duty, so in April 1951 I reenlisted for combat in Korea, which means I signed on for another three years of service just to get my ass out of the United States and into the war zone.

I didn't like the Army at all until I got a taste of combat in Korea. I advanced from a private first class to an infantry platoon sergeant while in Korea. More important, I learned what made men tick, and what combat was all about. For the first time in my military life, I felt completely at home. I could have asked for a more forgiving landscape than Korea, which was like no other place. We'd climb a hill, with great expectations of meeting the enemy, only to arrive at the top to see another, slightly larger hill looming. All the trees were stripped for firewood, and cold penetrated my bones. I was only twenty-one years old, so I handled the cold much better than later in life, but we Texans and Floridians in Korea were continuously cold. As far as wars go, Vietnam, with its insufferable humidity and constant heat, was much more to my liking.

Upon returning from Korea in December 1952, I entered Officers' Candidate School in Fort Benning. During the twelfth or thirteenth week, I contracted malaria and spent a week in the hospital. To return to OCS, I would have had to revert back to the eighth week, since my class was too far advanced for me to catch up with it. I refused this move and was sent to Germany as a sergeant first class and assigned to the 5th Infantry Division as a platoon sergeant. It was during my stay in Germany, sometime in 1953, that I read about Special Forces moving a unit to Bad Tolz, Germany. I began politicking for a transfer to SF, and I made a trip to Bad Tolz to see for myself. Once I learned what these fine men—the fittest and most committed group I had ever seen—were to become, I knew it was the only place for me. I immediately cranked an intertheater transfer and had it granted, to the 10th Special Forces in Bad Tolz. From the moment I joined those fine and fit men, I knew I was there to stay. It was, by far, the best move I ever made in my life. I might leave Special Forces, but Special Forces would never leave me.

So as I lay on the ground in the Bong Son rice paddy, I was forced to imagine my life without the Special Forces, without combat, without an enemy to fight. I didn't like the thoughts that raced through my head, so I shoved them out of my mind and went to work thinking about what it would take to get my body back together and back where it belonged, on the field of combat.

My journey to this unenviable position, with my body shot up in so many different areas, began in Okinawa at the beginning of March 1965. I was asked by Lieutenant Colonel Elmer Monger, the commander of Company C, 1st Special Forces Division, to assemble an A team—consisting of the toughest jungle fighters—to disrupt the enemy's movements in the Bong Son area, in the northeast section of Binh Dinh Province, along the South China Sea. At this time, Binh Dinh Province was completely controlled by the enemy, so I knew plenty of action would be coming our way. Captain Paris Davis, our excellent team leader, was assigned by group headquarters, and I was the man with the most combat experience. Our mission was to enter the area secretly, live there, build a Special Forces fighting A-Camp, and train locals to take the action to the NVA in his home. When Lieutenant Colonel Monger asked me to assemble this team, I accepted the mission with a crisp salute and the words "Roger that, sir." I have always believed this type of mission was my reason for being on this earth.

After we prepared for the mission in Okinawa, we traveled as a team to the Republic of Vietnam on a C-124 to Qui Nhon, the capital of Binh Dinh Province. There we picked up two unmarked U.S. Army trucks, painted jet black with no military markings whatsoever, for the eighty-kilometer trip north to Bong Son.*

Intelligence reports had alerted us to the heavy NVA presence in Bong Son, our new home away from home. One hallmark of a Special Forces A Team is its ability to get behind enemy lines and build a working camp from the ground up, using a bare minimum of supplies. So for us, this was nothing new. We chose a spot along the An Lao River (clear and fast-flowing at this time of year) that included a clearing that could be used as a landing strip. Our supply list began and ended with the following: one roll of concertina wire, a bunch of shovels, and a stack of sandbags. We started digging, working our asses off day and night. It was great work, and rewarding. We dug until we established smaller holes for individual fighting positions and foxholes, and several larger holes for our communications position and a headquarters bunker. We didn't know how long we would stay in the area, but we knew there was enough NVA activity to keep us busy.

We had a lot of work to do and not much time to do it. We would be receiving newly recruited mercenaries very soon. Our job was to train these men for combat versus the enemy, then conduct combat against the NVA that infiltrated

---

*Our team was assigned to 1st Special Forces in Okinawa, and on this mission we were on temporary duty (TDY) to the 5th Special Forces in Vietnam. Our bosses, the 5th Special Forces B team, were based in Qui Nhon. The B team, which might control three or four A teams at any given time, was responsible for our operational and logistical support.

our district area. Our plan was to engage the NVA in every direction for at least twenty kilometers surrounding our base.

Bong Son was strategically important; the NVA was using the port along the South China Sea to drop off soldiers from the north. They would land at night, in an area that was not a port but simply an available boat landing site, about eighteen kilometers to the east of Bong Son. They arrived by the hundreds in small motorized boats—wooden, flat-bottomed, bargelike boats that could maneuver through the sandbars. Despite the small size of these boats, the NVA piled as many as four hundred soldiers into each one, giving them the look of refugee boats. We didn't have satellites at that time so we had to rely on human intelligence to let us know what was happening.

Gathering intelligence is what my old friend Master Sergeant Anthony Duarte of Special Forces Delta Project did especially well. After leading a reconnaissance mission in the area, he confirmed the reports. He also told me, "These aren't local hire VCs. They're well organized and equipped NVA regulars with some Chinese among them." I passed this information along to Team Leader Davis and our control unit, an SF B team in Qui Nhon.

We sent a native Vietnamese speaker out to recruit mercenaries to aid our cause. In the official vernacular of the U.S. military, these mercenaries were called a Civilian Irregular Defense Group (CIDG). Staff Sergeant David Morgan, who had completed three tours in Vietnam, traveled with the recruiters, using money provided by a section of the CIA called Combined Studies Division. Morgan and the native speaker found the recruiting of raw, eager young South Vietnamese men pretty easy. We also went to the Bong Son district chief in an attempt to recruit. Of course, he was on the payroll also, an absolute must if we wanted to keep him on our side. I don't know if these mercenaries believed in our cause, but they would do the work as long as the pay was right. They were willing to train and had no trouble with the living arrangements. Most of them bounced around from one camp to another and did pretty well for themselves financially. They knew the risk, and they knew what was expected of them. That was good enough for us.

We recruited around one hundred or so Viets from the lowlands, then moved them into our area and began the work of supplying them with clothing and weapons and training them to fight. Meanwhile, construction of the camp, supervised by Morgan and carried out by a civilian Vietnamese work crew, continued. Every Special Forces man was working twenty hours a day, and the work was not without risk. The NVA sent their love to us often, in the form of small-arms and high-angled harassment via 82 mm mortar fire, plus an occasional B-40 rocket-propelled grenade (RPG) directed into our new positions.

We knew the NVA monitored our movements, but we didn't give a good goddamn. We had a mission. Our local intelligence kept us informed of activities that might appear to be a concentrated attack on the fledgling SF camp, but we received no threats of imminent overrun. We also sent out short-range patrols nightly, to keep tabs on enemy movement.

One week before the proposed raid, five of us set out on a reconnaissance mission. In the early predawn light, we moved approximately fifteen kilometers to the east, toward the coastline of the South China Sea. Once there, we stood in a cemetery on a bluff about three hundred meters from an enemy camp and watched about fifteen NVA soldiers bullshitting and walking around. We could see three bamboo sleeping hooches behind the milling soldiers. As we watched I told Morgan and Staff Sergeant Ronald Wingo, "These guys look a little too comfortable. We've got to raid this fucking place, take out those huts, and stir these bastards up." We chose the high ground of the cemetery as our rallying point after the raid. Our small recon group was spotted by a few of the NVA soldiers, who promptly opened fire. We returned their fire but ended up on the deck of that cemetery, lying on top of one another above the graves of dead Vietnamese farmers.

Wingo and I used one of the gravestones as cover against the oncoming fire. Lying prone on the deck, we called in a few air strikes to let the enemy know we meant business. In the meantime, the NVA machine gun chipped away at the headstone. Between ricochets, I said to Wingo,

"Ronnie, I know exactly what the fucker in this grave is thinking."

Wingo was not amused by our predicament and wasn't interested in my using this moment to theorize on the thoughts of some dead Viet.

"Goddamnit, Billy, I don't care what the fuck he's thinking. Fuck him. We're in some deep shit here."

The headstone was getting smaller and smaller.

"Well, Ronnie," I said, "this fucker is saying to himself, 'I'm sure glad I'm down here and those two dumb asses are up there getting their asses kicked.' "

I laughed, but Wingo didn't see the humor of the moment until later, after we made it out of the mess intact. That's one of my rules of combat: Sometimes you have to laugh in the face of horrible situations. Soldiers without a sense of humor were eaten up from the inside; those who could laugh were more likely to retain their sanity.

Our reconnaissance confirmed the enemy was obviously prevalent and not too concerned with our presence. We decided then and there, by God, to plan the raid for the following week. We would change their attitudes about us by returning with our team and newly trained troops to kick their asses.

This raid was well considered. We had scouted it. We had intelligence on the enemy camp. We were strong and confident. We would attack with the Special Forces hallmarks—speed, secrecy, and surprise. As it turned out, despite our considered planning we were still sadly unprepared for what the NVA had in store for us.

We marched through the night and early morning of June 18, 1965. By 0430 it was 85 degrees and clear—combat weather—as I led the formation through the thick air along the An Lao River. There were ninety of us—four U.S. Special Forces and eighty-six of our native South Vietnamese mercenaries. We had left our camp at 2300 the previous day to walk the seventeen kilometers through the trails along the river to the spot we knew the NVA lay sleeping in their bamboo hooches.

The other three Special Forces men on the mission were Captain Paris Davis, Staff Sergeant David Morgan, and our medic, Sergeant Robert Brown. Davis was the commander of the team, but since this was his first combat mission, I was put in charge of leading the raid. Nobody is ever completely combat-ready on their first mission, but that changed for both Brown and Davis over the course of June 18. Brown was a tough kid with a quick smile and an All-American look about him. Davis, from Washington, D.C., was a blue-eyed black man with a confident air and a good mind for combat. Morgan was battle tested and savvy, a hardened veteran unafraid of fierce combat or hard work. Very few people in the world could build a Special Forces camp from the dirt up the way David Morgan could. I hand-picked these men for this battle, choosing those I felt were best prepared to think quickly and intelligently if the situation turned nasty. In Special Forces, nasty is a way of life.

The rookies may have been scared, they may have been nervous, but this isn't like the movies. These men went through Special Forces training, so the idea of combat wasn't anything new to them. You have to remember: This was the reason we were here. People unfamiliar with war and warriors have a hard time with this. It's not like we sat around a campfire the night before and talked about our feelings and fears. It doesn't work that way. Here's how we prepared: I told them, "You're going to be fine. Just do what you know how to do."

This particular job meant going into an enemy camp, taking out as many as possible, then leaving before the survivors could respond. Like every Special Forces mission, the raid was daring and dangerous. We had very little in the way of protection and nothing in the way of backup. It wasn't anything new to me; I had done it before and would do it again, many, many times. This was our job. Some people work in an office, some people run around the jungle in a clandestine manner and kill. We set out hoping this would be just another day at the office.

We carried the tools of our trade: an M-16 and twenty-five magazines of .223-caliber ammunition slung over our green jungle fatigues; ten frag grenades; two white phosphorous grenades; two smoke grenades. Each of our twenty-five M-16 magazines consisted of twenty rounds, giving us five hundred rounds each. We were loaded. We also carried Special Forces rescue gear, consisting of a mirror, a red flare, a red emergency panel, and a compass. You hope you don't have to use those. I even grabbed a few handfuls of hard candy, for energy.

Our communications equipment, a PRC-25 FM radio, was operated by one of the indigenous "irregulars" to keep in contact with the Forward Air Control (FAC) aircraft. It was a small Air Force Cessna O1-E piloted by a USAF pilot with my SF team demo man, Sergeant Ronald Dies, riding shotgun. As the day wore on, I would wish like hell I had carried a radio myself.

BONG SON 1965 — CONSTRUCTED BY SF

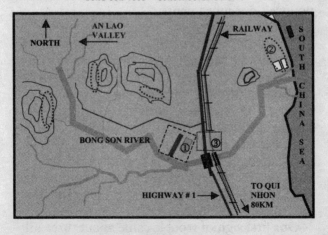

①	Bong Son SF camp
⋯⋯	NVA-owned
②	NVA-stronghold—site of raid

Our intention was to walk right into Charlie's home and kick his sorry ass out of the Bong Son area, once and for all. We would destroy his camp and administer a beating his ancestors could feel. And then we'd haul ass out of there before Charlie could recover.

We mustered near midnight on the night of June 17 and departed the Bong Son Special Forces camp at 0100. We knew what lay ahead: a seventeen-kilometer march along the An Lao River over narrow single-track trails that snaked in and out of the riverbed. We walked silently and swiftly as we cleared the small village of Bong Son. Not even the ever-present dogs announced our movements.

I walked toward the front of the formation, just behind the point man, with thirty of our mercenaries. Morgan and Brown walked near the front, also. Paris Davis stayed toward the rear and controlled the remaining fifty-six of our troops. This was the first time we had employed this newly trained CIDG unit in combat. They were not that good, frankly, but they had potential. I thought a quick, successful raid would do wonders for their confidence and competence. That was the hope—in and out, congratulations all around.

Our first sign of trouble came about three kilometers outside our own SF camp, along the riverbank. We came upon an NVA medic sleeping along the trail. This poor bastard, with a Chinese pistol in his holster, was supposed to be standing guard for the camp, but he was fast asleep and therefore absolutely no good to his side. In my twenty-seven months in the region, I had dealt only with local-hire Viet Cong, but here was my first direct contact with an NVA regular. When we captured him, I couldn't believe the fine medical gear he carried. "Look at what this son of a bitch has on him," I whispered to Morgan. He shook his head and whistled silently. The medic's bag was filled with top-notch items, with the same life-saving equipment our medics carried. This guy's equipment was sobering and slightly alarming; it made me think we were in for more than we expected. He had brand-new gear, brand-new boots, brand-new pith helmet. He never got a chance to use it, though—we took the bag and everything in it, then I had one of the mercenaries slit his throat. This was all done silently and efficiently, at about 0200 that morning. For all intents and purposes, the raid had begun.

Further along the trail, maybe two kilometers before the NVA camp, we came across a man and a woman in a large fenced-in area, cooking breakfast for the soldiers. It was 4:30 in the morning, and they were hard at work. Pots of food were boiling over the fire, and the man was helping her break up firewood when they looked up and saw us bearing down on them. I was the first to come upon them, and the first thing I saw was the pistol strapped to the man's right hip. Before he could react I was on him, with a knife to the throat, and then I discovered yet another

surprise: His companion was one tough Vietnamese lady. She picked up a stick and started
whacking me upside the head with it, beating the
shit out of me. She wouldn't stop with the fucking stick, and she was loud and getting louder.
This wasn't part of the plan. We couldn't afford
to have this woman wake up the whole damned
camp, so we had to subdue her and kill her the
same way we killed her cooking partner. Since
we hadn't fired a round, the quietness prevailed.
The soldiers in the camp remained asleep and
oblivious to our presence.

That, as it turned out, changed soon enough.

We walked stealthily, and maybe a little faster,
after our encounters on the trail. Word had been
passed down the line, to Davis at the rear of our
formation, that the enemy was within meters.
Our men moved without a sound. As we reached
the first hooch just off the wide, well-worn trail
on the edge of the jungle canopy, each man could
hear his own heartbeat and nothing else. The
adrenaline running through my body was a
force, accentuating my senses and giving me a
feeling of strength and invincibility.

Dozens of uniformed NVA soldiers were
sleeping within these huts. The roof of each hut
was about three feet tall, covered with branches
and leaves from the surrounding jungle. The
NVA slept in their uniforms, uncovered, in the
open air, three to a bare wooden platform just
above the dirt. The platforms were approximately four feet wide, no wider than a queen-
sized bed. Each hut was about forty feet long and

forty feet wide, and about sixty men slept inside each.

We passed messages by prearranged hand and arm signals, beginning with me and traveling through Morgan to Brown. We broke up into three groups, one for each building. I took one, Morgan took one, and Brown the third. We each had seven or eight of the mercenaries with us. Davis was controlling the main body of soldiers and had not yet reached the sleeping huts.

Our plan was to wait for the full force to arrive before beginning the assault, but after the incidents with the NVA medic and the cooks, it seemed too risky to wait for Davis. As I saw it, delay meant certain discovery. We needed to attack and kill these sleeping beauties.

We positioned ourselves in the entrances of the huts, making sure our weapons were ready, grenades in hand. At that moment, the moment before combat, I didn't hear a thing. My mind was silent, my body calm. When I gave the sign, we worked in unison. We announced our presence with a couple of grenades for each hooch, followed by round after round of M-16 fire. We sprayed bullets around those huts like firemen dousing flames. We fired and reloaded. Our empty magazines dropped to the ground. The slaughter was on, and many NVA died in place while others tried to escape through the back of the huts. Not many made it.

The sound . . . I remember the sound. After so much concentrated silence, the sound of gunfire tore apart the predawn calm. And then there were the screams. Panic has a distinct sound, and this was it. This was the sound of men with nowhere to go and no way to get there. The human instinct is fight-or-flight, and these guys weren't in a position to fight. They were trying their damnedest to make flight a reality, but it was futile.

As this took place, I scanned the area. I couldn't help but notice the NVA's weapons as they leaned against the hut posts and lay strewn about the jungle floor. The hut—as well as the other two—was filled with weapons and gear. These guys were equipped. They were ready. A thought kept racing through my mind—*this is a first-class outfit; I don't like the look of this*. Not many U.S. soldiers had seen NVA regulars at this early point in the war, and I was shocked, realizing the well-equipped medic along the trail was not an exception.

But as we looked over the spoils of our victory, euphoria ran right the hell over any misgivings I might have had. We loaded up our local mercenaries with the gear from the dead NVA—Russian RPKs and AK-47s; radio equipment with advanced communications and intercept. My heart rate was blazing. Speed, secrecy, and surprise—damn, this was Special Forces work at its best: a well-planned mission run to perfection.

Killing the enemy causes a man's adrenaline to run wild, even wilder than the preraid buzz. I felt like Superman, able to perform actions far beyond the normal human capacity. We all felt like nothing could stop us. Our entire force, Davis and the remaining troops now included, gathered and counted the loot for about fifteen minutes. As we were still loading up and celebrating, patting ourselves on the back, we heard a sound that pierced my heart. Out of the depths of the jungle, deeper still from where we were, the call of bugles rolled out in waves.

The congratulating stopped.

That sound meant a raid was about to become a fierce, bloody battle.

That sound meant deep shit.

The hair on my arms stood on end. I turned to David Morgan and said, "David, our asses are in trouble. It's time to vamoose."

Morgan wasn't aware of all the implications of that sound, but I had heard it in Korea. Bugles meant Chinese soldiers. Chinese regulars.

And a pending counterattack.

We didn't have any idea that a string of dozens of bamboo huts stretched deep into the jungle, and we had no idea that each one was filled side to side and front to back with the best soldiers the NVA had to offer. The NVA built their huts deep under the jungle canopy, and the roofs of these huts can't be seen from above. In fact, they can't be seen until you're right next to them. We had no spotting devices in 1965. We had only men, men who would go into that jungle and drag the NVA out by their necks, knock out a few teeth with the butt of your weapon, and

convince the enemy he was a dead man if he
didn't cooperate. Pure battle is what it was.

The three huts we hit were just the beginning.
Duarte had only two other men on his Delta
Force recon team, and later we were informed
that more than ten boats had unloaded soldiers
to that camp in the previous few days. Close to
four thousand new soldiers were sleeping in
those jungle huts, and every last one of them
knew what had happened that morning to the
huts we visited. We conducted the Bong Son raid
based on information from headquarters, along
with our own reconnaissance. Nothing we saw
or heard during our recon mission or on the
morning of the attack indicated a camp of this
magnitude.

With the bugles ringing in our ears, Morgan
and I ran into the jungle and past several more
huts, tossing in hand grenades and completely
expending our ammo; we slaughtered row after
row of enemy soldiers as they attempted to mo-
bilize.

Seemingly all around us we heard the clatter of
men preparing for war: the rapid footfalls of thousands of boots on dirt, the readying of weapons,
the yells of surprise. We had stirred up more than
we could handle. It was time to get out of there,
and fast.

The shots started coming, green NVA tracers
from the Russian RPKs, as the soldiers began to
line up in formation. I fired two handheld red
flares, signaling our forces to withdraw to the
high ground to the west, the same cemetery loca-

tion that had served as our previous observation post.

I ran due west toward that rallying point, sprinting like a scraggly assed ape, every one of my senses on high alert. The enemy was screaming over the crack of AK rounds buzzing around my ears. When a bullet is meant for you, and you alone, you will know by the way it cracks right beside your ear. That sound is like no other sound you have ever heard, or ever will.

They started pounding our asses and didn't stop. To reach our rallying point at the cemetery, we had to cross a wide rice paddy. Fortunately for us, it was a few months away from the monsoon season, so the floor of the paddy was hard and dry. I could hear the low clumps of grass schuss under my jungle boots as I raced through it. If it had been flooded, and we had to splatter our way across that paddy, there would have been no hope for any of us. We didn't have much going for us, but we could run. You take what you can get.

I was out—out of ammunition, out of grenades, nearly out of hope. And no more than thirty meters across the paddy a green tracer from one of the Russian RPKs drilled me in the right knee. The impact knocked me off my feet, forcing me to crawl. The pain was intense. I pulled myself toward a small irrigation levee, a six-inch berm of dirt that ran the width of the paddy. The rich loam of thousands of years of agriculture rose off the ground and into my nostrils as I used the small levee as a shield against the incessant fire. I continued to crawl west, my body pressed against the edge, pointed toward safety.

Although I didn't know it at the time, my young SF medic Brown had already been hit in the head by an AK bullet. He would survive the battle but would die later on the operating table of a MASH unit. Dozens of our mercenaries were strewn about the battlefield. Many of them went down before they had a chance to attempt an escape. Others ran at the first sign of trouble and were picked off like clay pigeons as they ran howling across that paddy.

Hard to believe I was one of the lucky ones, with my knee shot to hell and the tracers flying inches over my head. I crawled along that levee about forty meters to a dug-out pit in the middle of the paddy. It looked like a good foxhole to me, especially given my predicament, and when I reached it, I felt a brief flash of relief as I flopped onto its muddy floor.

This pit was about four feet deep and twenty feet long. I rolled over and looked up and . . . *holy shit* . . . I was staring directly into both barrels of a water buffalo's flaring red nostrils. He was not a welcome sight. This mean old boy was staring at me like I'd invaded his living room, which I guess I had. These animals were prized for their work in the rice fields, and the pits were dug out to give them a place to rest and cool off from the relentless heat. He was ready to hook me with his horns and eviscerate my scrawny ass—to finish the job the NVA had started, I guess—but the old son of a bitch couldn't get his legs moving. He was stuck too far into the mud to make any kind of frontal assault. This was a new twist to a bad predicament: chased by bullets into a muddy hole with a water buffalo, with the wild screams of the enemy getting closer every second. I looked that bastard dead in the eye and said, "You big son of a bitch, move over." He was snorting at me and butting his head at me, but he couldn't reach me. I'm on my knees, I'm in pain, and the enemy's closing in. And what am I doing? Talking to a fucking water buffalo.

"You ready to fight?" I asked straight into his snorting snout. "Let's go, big boy."

I heard a commotion above me, a frantic disruption of earth, and here came Morgan into the pit with me. He looked at me, then at the water buffalo, then back at me.

War is hell and all that, but sometimes it's pretty damned comical. We would have loved to stay in there and rest, but we knew the protection was short-lived. The NVA knew we were in there, and they were fixing to charge across that paddy. We either had to get out of that hole or be shot to pieces lying there next to that dumb-ass water buffalo. I unfolded my two-foot-by-two-foot red canvas emergency panel from my pocket and dropped it in the hole, hoping our FAC commander flying above would see it and call in the air strikes. I told Morgan, "Last chance, buddy—let's get the hell out of this place."

We had to go it alone. The ground fire from the NVA was too heavy for Morgan to attempt to save himself and rescue me at the same time. Our best chance, our only chance, was for Morgan to work his way through the hail of fire toward the rallying point, where he could call for air strikes.

He looked at me before he climbed out, as if giving me a chance to change my mind.

"Go on," I said. "Get us some help."

"We'll get you, Billy," he said.

Morgan jumped out of the pit and weaved his way in a low crouch toward the rallying point.

I was on my own. Soaking wet from the waist down and covered with mud, I could feel the leeches as they ate their way into the wound on my knee. Enemy fire pounded over the hole so persistently it became background noise. I sent a good thought Morgan's way as I dragged my leg up out of the buffalo's den and crawled in the direction of the withdrawal area, defenseless but still determined.

I crawled less than five meters and took another hit. This one was the worst: The tracer round pierced straight through the sole of my right jungle boot and shredded the metatarsals of the big toe and the two adjacent toes, tearing through flesh and bone and exiting above my ankle.

Pain. Oh my God, the pain. There aren't words for this kind of pain.

The thought occurred for the first time: *Damn, my military career is finished. I'll never see combat again.*

I went facedown and saw the tracer sitting on the brown-red earth a few inches from my nose,

AIR MAP OF THE BATTLE AREA

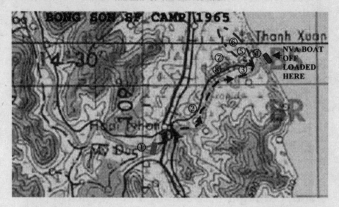

① 0100 SF Camp Bong Son—1965 loc: N 14° 29′ 30″ E 109° 01′ 00′ 15 Apr 65 Starting point of raid operation

② 0300 Capture NVA, asleep on guard duty, full medical kit, new uniform, good gear, and a Chicom pistol

③ 0500 Capture one male / one female cooking food, both armed with AK-47s

④ 0515 Kill 60+ armed NVA asleep in three thatch hootches, who had intentions of killing U.S. forces

⑤ 0600 Kill an additional 100+ NVA as our raiding force moved through the target area

⑥ 0645 MSG Waugh hears Chinese bugles calling for counterattack by the NVA. Waugh signals withdrawal.

⑦ 0730 Friendlies suffer heavy casualties as the enemy springs an ambush as friendlies cross a rice paddy. Waugh SWIA, Brown KIA, 80 friendlies SWIA/KIA. Area of the water buffalo wallow—battle lasts six hours.

⑧ Recovery area, where med-EVAC helicopters came for WIA, and resupplied ammo to beleaguered friendlies

green and spinning like a piece of fireworks. I knew those green tracers glowed for ten seconds, so as I watched it go dark after a few seconds, I calculated the distance in my head: The enemy couldn't have been more than fifty meters behind me.

I kept going, somehow, willing myself to crawl inch by inch toward the bamboo stand that lay about twenty meters ahead. My right leg was useless, and I dragged it and its exposed bones behind me as the sweat poured off my forehead and more leeches made their way into my fresh wounds. I couldn't hear a word from any of my people.

The green from the tracers continued to stream over me, like a goddamn light show.

I just told myself to keep going, keep going. I couldn't fight, but I could still flee. It took me ten minutes or so, but I traveled those twenty meters and dragged myself behind that thick bamboo stand, putting it between me and the enemy.

My heart pounded as I struggled to catch my breath. Sweat dripped off me, as much from pain as exertion.

This was as far as I could go. The cemetery lay ahead of me and above me, maybe forty meters west and fifty meters high. I couldn't get there, not through all that gunfire, not in this condition.

As the NVA bullets whipped and ricocheted through the bamboo, I took stock of my body. I could see the bones from my right foot and ankle, white as snow, sitting on top of my right shin. My foot had been torn from its moorings, and blood was everywhere. My right knee was filled with shrapnel and covered with leeches. I had also been shot in the left wrist, the bullet shooting off my watch. The reasonable part of my brain kept saying *I'll never walk again,* but the rest of me said *Somehow, some way, I'll be back here someday.* I got into my rucksack and

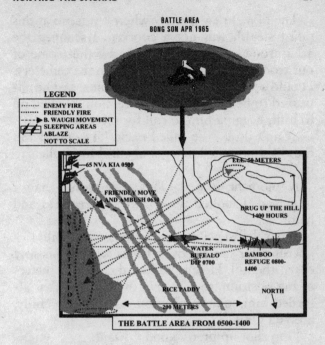

**BATTLE AREA
BONG SON APR 1965**

LEGEND

	ENEMY FIRE
	FRIENDLY FIRE
	B. WAUGH MOVEMENT
	SLEEPING AREAS
	ABLAZE
	NOT TO SCALE

65 NVA KIA 0500

FRIENDLY MOVE
AND AMBUSH 0630

ELE. 50 METERS

DRUG UP THE HILL
1400 HOURS

NVA BATTALION

WATER
BUFFALO
DIP 0700

BAMBOO
REFUGE 0800-
1400

RICE PADDY

NORTH

200 METERS

THE BATTLE AREA FROM 0500-1400

fumbled around until I managed to give myself three shots of morphine with field syringes. I lay there, waiting for it to take effect, but it was no use. There was no messing with this kind of pain. Those three shots of morphine were like a couple of half-filled sandbags against a flood.

I have to say this situation disheartened me quite a bit, and it made me understand that whatever was going to happen to me was going to happen to me right there, in that little bamboo cover in the middle of that rice paddy. My fate was right here.

The NVA knew exactly where I was, so at this point stealth was not a concern. I scanned the area around my position and saw that some of our troops were returning fire from the cemetery. I could see Davis at the rallying point, and he appeared to be looking in my direction. I yelled up to him, begging him to call for tactical air.

"Can't do it," he called back. "The goddamn radio's shot to hell. It's got about thirty bullets in it."

A moment passed. This was worse than I imagined. I looked up at Davis, speechless and helpless.

Davis yelled, "I'm coming to get you, Billy!"

Using our fire as cover, Davis worked his way down the hill. He got within about three meters of my position, to a spot in the bamboo that afforded him some protection, and asked, "Billy, are you OK?"

"I'm shot up pretty good," I said.

He asked me if I had any commo, and I told him I had nothing, not a damned thing. I looked straight into Davis's blue eyes. I could tell he was feeling the same thing I was feeling: dread. My words came out in a near whisper when I said, "If we weren't already, we're damned sure in trouble now, my friend."

Before I could finish the sentence, Davis got shot across his right hand. The ends of his fingers were clipped clean off, and he was screaming and cursing. His hand spurted blood.

"Goddamnit, I can't even shoot now," he yelled. "I'm right-handed."

I'm lying there, unable to move, hoping Davis can get out of here alive, and I said, "Well, hell, Paris—you're gonna have to pull the trigger with your left hand."

Any hope of being rescued disappeared the moment the bullet sheared Davis's fingers. Holding his hand, he worked his way back to the cemetery's high ground in an attempt to signal to the FAC controller that our unit required both medical evacuations and communications equipment. Having lost radio contact, our communication was limited to emergency panels and mirrors. Within minutes of Davis returning to the rallying point, the U.S. Air Force pilot in the two-seat OE-1 above assessed the situation and concluded that the NVA was controlling the battlefield. He saw the Christmas-party exchange— a whole lot of NVA green tracers going one way, not many U.S. red tracers going the other—and knew we were in deep trouble. He called through his Airborne Command that U.S. troops were in trouble, and emergency Tac Air was required.

It wasn't long before we could hear the sweet sound of navy F-8s and USAF F-4C Phantom jets coming in low and fast over that battlefield. The USAF FAC was a master at employment of these fast movers, and he put these fighters to work, initiating low-level napalm and strafing. These birds worked in pairs, with one covering the other, as they swooped in to drop their napalm and hard bombs onto the rice paddy and to the east where the NVA had massed.

The Battle of Bong Son marked the carrier-

borne navy fighters' first tactical air strikes in the
Vietnam War, and it couldn't have come at a better time. They came off the USS *Washington,* a
carrier in the South China Sea, and I'd never seen
a more welcome sight. Those fighters kept the
NVA from advancing toward those of us who
couldn't help ourselves. The napalm landed close
to me, close enough that I felt its heat and
smelled the kerosene. The high-explosive (HE)
bombs, though cracking and ear-splitting, were
music to my ears.

Damn, I love those air strikes.

Through it all the NVA was still screaming,
and still blowing those hellish bugles, advancing
to within twenty-five meters of me. The air
strikes impeded their progress, but there were so
many of them they just kept coming.

They were gaining on me, and I was trapped,
boxed in by the gunfire, the bamboo, and my
own quickening mortality. It didn't look good,
so I figured this was probably a good time to
have a few words with The Man upstairs. Now
or never, right? I tuned in my brain to God and
made a few promises. "Listen," I said out loud,
"I owe you a lot. Tell you what: You get me out
of this fix and you've got me. I am yours."

My prayers were answered, but not in a here-
comes-the-cavalry way. I wasn't rescued, at least
not yet. But as I hid there, trying not to look at my
exposed bones and the leeches that were enjoying
the situation, I took another bullet, this time
across the right side of my forehead. I don't know
for sure, but I believe the bullet ricocheted off the

bamboo before striking me. It sliced in and out of a two-inch section of my forehead, and it immediately started to bleed like an open faucet.

It sounds like the punch line to a bad joke, but you know it's a bad day when the best thing about it is getting shot in the head.

So how did this save my life? It saved my life because it knocked me cold and made me appear awfully damned dead. I wasn't far from it, but the combination of the exposed bones, the leeches, the wounds to the knee, wrist, and forehead—if you came upon me on a battlefield, what conclusion would you reach?

The sun was beating down hard on me from overhead. I woke up dreamily, my brain a bit fuzzed. I wasn't coherent enough to calculate with any specificity, but the location of the sun meant I'd been in this position for at least eight hours.

All that mud had baked on me like a crust. The leeches were everywhere. The bones on my leg were sun-baked. The dried blood on my forehead made it tough to see, but I didn't need my eyes to understand I was naked. They'd come across that paddy and stripped me of my clothes, my Rolex watch, my gear—everything. The only thing that saved me was The Man's answer to my prayers—the shot to the forehead. As I write this, I suppose I still owe The Man on some of those promises.

I heard a helicopter. The sound gradually worked its way into my conscious mind, bring-

ing me my first brush with hope since that first
tracer struck my knee almost ten hours before.
The chopper had arrived to deliver communications gear. I tried to crawl, but it was no use.
Though the navy Phantoms were still pounding
the NVA, the NVA was still firing at will.

One of our Special Forces men, Sergeant First
Class John Reinburg, landed with one of the incoming helicopters and crawled below the enemy
fire to the bamboo area. Reinburg was the
heavy-weapons leader for our SF team, and he
had come to assist in the rescue. I looked up at
him in his U.S. jungle fatigues, not sure whether
to believe he was really there.

"Listen, Billy," Reinburg said. "We're gonna
get you out of this frigging mess."

I was fine with that, and my spirits were
boosted quite a bit, but I wasn't sure how it was
going to be accomplished. By now a HU-1D helicopter had landed on the reverse slope of the
rally point. We were still forty or so meters from
the base of that hill. From there we'd have to
climb fifty meters without getting shot, then
make it to the helicopter.

Reinburg grabbed me from behind, holding
me under my arms, and started dragging me toward the hill. We got out from behind the bamboo cover. The tracers kept coming, but there
were fewer of them than I remembered. I could
hear my Special Forces guys up above, rooting
us on. At some point Davis came alongside
Reinburg and they kept pulling, six feet at a

time, toward the base of that hill. "We're gonna make it, Billy. We're going to make it," they kept saying.

I was starting to believe them. They kept yanking me up that hill, and the whole time the bones in my right foot and ankle were flopping around beneath me. I was in so much pain I felt sick to my stomach. Farther and farther we climbed, to the point where I could look over my shoulder and see the top of that damned hill. Was this really happening? After the events of that day, I wasn't sure I could trust anything. I wasn't sure whether this was real or imagined.

As we neared the crest of the hill, Reinburg grabbed me tight and gave one last pull to drag me the last few feet to the top. As my body flopped on the top of the hill, Reinburg stood up over me and just then took a bullet to the left breast, just above the heart. As he started to go down another bullet hit him about six inches lower. The two bullets bracketed his heart and severed both lungs. Now we were both down, and in that instant Reinburg went from being strong and heroic to being in worse shape than me.

I put aside the pain and, with Davis's help, started crawling toward the copters. I knew if I stayed low I was relatively safe, and for the first time in a while I started to believe survival was a distinct possibility. All I could think was *Crawl, you son of a bitch, crawl*—until I got close enough to be loaded onto the helicopter. As I looked at Davis, Morgan, and the few CIDG

who were among the survivors who made it to
the high ground, I knew we needed to get some
automatic weapons out here to stave off the
NVA. Otherwise, they would surely overrun this
hill between air strikes.

Davis helped me into the helicopter, and I
owed him and the wounded Reinburg a debt of
gratitude for sticking with me and helping me
get out of that hopeless mess.

As I was placed on the floor of the helicopter,
one of the gunners took a green tracer round to
his left arm. The impact nearly sheared off his
arm at the elbow. Half his arm hung there, by a
thin strand of cartilage, while the gunner stared
in wide-eyed shock, unable to respond. They
piled the bodies beside me and below me and
above me, and when the floor of the helicopter
was covered with bodies they began stacking
them on top of my naked body. Dried blood and
burnt skin and exposed bones filled the helicopter, along with the cries of the wounded. I didn't
care. Dignity was not an issue. The cries merged
into one sound, a constant wail of pain.

I wanted to get the hell out of that place and
start healing. The way I saw it, I needed to get
out before I could get back. One step at a time.

When that chopper took off, I was off the battlefield, but my mind never stopped working. While
lying on the deck of the chopper, I managed to
make some room for myself and call my radio
operator, Sergeant First Class Kenneth Bates, who
had remained in the camp to send and receive ra-

dio messages. I told him to get at least two .30-caliber machine guns out on the next chopper on its way to the rallying point. I told him to get ammunition and two PRC-25 portable radios to Davis and the other lads before they were overrun. Bates gave me a thumbs-up as we departed for the MASH facility in Qui Nhon.

Bong Son was a bloody, savage battle, to that point the largest land battle of the Vietnam War. Brown died of the head wound he received early in the battle, and the other three of us SF men—Morgan, Davis, and I—made it out through a combination of luck and determination. Just fifteen of our eighty-six mercenaries made it out alive, meaning our raid had an 80 percent mortality rate. Ninety went in, eighteen came out, and everyone who made it out left in the supine position.

A Ranger unit was finally inserted into the target area in the late afternoon of that fateful day, and it estimated the NVA lost more than six hundred troops in the Battle of Bong Son. NVA soldiers were killed along the trail and in the huts and on the rice paddy, cut down by the force of the Navy F-8s and the USAF F-4C Phantoms tactical air strike aircraft.

As that bird lifted off the landing zone above that rice paddy with my battered body on board, I looked down and watched those green tracers continue to fly over the dead bodies on that battlefield, and I made a vow right then and there: I might be shot to shit and broken in a whole bunch of places, but I couldn't let my military career end like this. I simply couldn't.

I looked down as that rice paddy receded beneath me, and I made a vow.

There was a war to be fought, the only good war we had at the time.

And I'd be there to fight it.

CHAPTER 2

The people around me told me how bad I looked, and they did it without saying a word. When I was removed from the military ambulance at the 8th Field Hospital in Nha Trang, I was greeted by a nurse whose first reaction was a startled shriek. She began cursing and turned away. I thought she might vomit, right there in front of me. She shook her head, looked over her shoulder, and said, "I can handle anything in this war but those goddamned leeches."

The other nurses wore the same look of shock and revulsion. Their reactions didn't fit with my mood. I was giddy, happy to be alive, high on Demerol, morphine, and massive blood loss. On the flight from Qui Nhon, I joked like a crazed person with the flight crew of the Army C-7 Caribou. The two crew members kept looking at me, shaking their heads like I had dropped in from another planet. They had good reason: My clothes were absent; I was wrapped in a sheet; I was playing host to a colony of leeches; and I was laughing and joking like a man possessed. It would have taken a lot to insult me in that condition. I was alive, though, people—*alive*. I had just spent how long—ten hours?—in a living hell, facing an enemy that was both pissed off and well armed, and I made it. I fucking made it. A few leeches, what the hell? Couldn't these folks tell by looking at me how long the odds were? Couldn't they look at me and see what I'd seen? At that moment, that brief moment of narcotized elation, I thought everybody should be happy to see me.

That was my world. In their world, the real world, the people who came into contact with me saw only a man covered with leeches, mud, and blood. They saw the bones of my right foot lying atop the instep of that foot. They saw blood from a gunshot wound to my left ankle, from a gunshot wound to my right knee, from a gunshot wound to my left wrist, from a gunshot wound to the head.

In the field hospital, which resembled the television MASH unit, the doctor took one look at me and told the nurses to prep me for debridement. I was given a shot of sodium pentothal and away I went. Gone. Even more gone than I already was. My bloodstream—what was left of it, anyway—was awash in narcotics. For the next eleven months, eleven months of tortuous rehabilitation and unending frustration, elation would be a difficult emotion to recapture.

*   *   *

I don't know how long I stayed in the 8th Field Hospital. My best guess is somewhere between five and eight days. When I awoke from the anesthesia, I discovered I was in a hospital ward with two other patients. Several doctors were administering to those patients, and I had the wherewithal to recognize the patient to my immediate right: Reinburg, the heavy-weapons man from my team who had been shot twice in the chest after pulling me from the bamboo stand to the high ground. He was screaming uncontrollably.

As I lay there, listening to Reinburg howl and cry from pain, I still couldn't quite believe I had made it this far. Twenty-four hours before, it didn't seem even remotely possible. Amid my reverie, the Demerol began to take effect. Reinburg continued his harrowing screams. They began to fade, their horrific sound receding into the distance, and I drifted off.

In and out. Back and forth. Asleep, awake, somewhere in-between. That was my routine at the 8th Field Hospital. Five days, seven days, eight days—I don't know how long the pattern lasted. At least half the time I was there was spent in a drug-induced fog. Most of the rest of the time I was asleep. I learned to shout for Demerol whenever the pain exceeded my personal threshold, which occurred several times a day. Days and nights merged in a haze of sameness. My life was reduced to two sensations: pain, and the relief of pain. I had no idea what the hell was going on around me. I would drift away for four hours, then be jarred awake through the

sheer force of the pain. I would shout for the
Demerol, then drift back into the fog. Lucid moments were few. I do recall on one of the days I
was in the 8th Field Hospital, General William
Westmoreland pinned a Purple Heart on my pa-
jamas and said, "Master Sergeant Waugh, this is
your sixth Purple Heart, and your country and I
thank you for your hard work." I looked at
General Westmoreland with a stupid look on
my face and muttered some words of gratitude.

On my last day in the field hospital, I was
placed on a litter and prepared for departure.
The pain was ever-present, diluted by my constant Demerol haze. The nurses wished me well
on my departure. The one who had greeted me
with a scream the day I arrived told me she had
removed more than thirty leeches from my body
on that first day.

She looked at me like she might never get over
it.

My wounds sent me back to the United States for
the first time in five years. I had been out of the
country nonstop with 5th Special Forces for that
long, and I had no intention of returning this
soon. But I was transferred to Walter Reed Military Hospital in Washington, D.C. My right
foot, ankle, and leg were the most serious
wounds, but at this time my left ankle moved to
the top of the list. That gunshot wound was horribly infected and particularly ugly—the bullet
had clipped the ankle bones and ripped about
eight inches of skin up from the ankle. There

were twenty to twenty-five wire stitches holding the lower part of my leg together, and those stitches were straining under the pressure of my rapidly swelling flesh.

The doctors at Walter Reed were the best the Army had to offer in the area of body repair. Repair was what I needed, and pronto, but I wasn't encouraged when I discovered I would be placed in the lower-extremity amputee ward of Walter Reed. That gave me an indication of the challenge I faced, and the work that had to be done to make sure I kept each of my limbs connected to my body.

My left foot and ankle responded to antibiotics, eventually easing the pressure on the metal stitches. But now my right foot and ankle had become infected. No matter what course of action the doctors prescribed, these wounds simply wouldn't heal. I needed to get better, I was determined to get better, and I *had* to get better. A return to the war zone was absolutely necessary, the way I saw it, and the delay caused by the slow-healing wounds irked me no end.

My stay in Walter Reed wasn't all misery, though. While there I was awarded the Silver Star for valor for my work in the Bong Son raid. I'm not much for awards—they aren't my motivation—but I have to admit I was proud to receive the country's third-highest military honor.

For the most part, my stay at Walter Reed was filled with boredom and gallows humor. One day, with a long-leg cast on my right leg and a short-leg cast on my left, I was in a Washington,

D.C., dance club with an Army sergeant who had lost his right foot above the ankle. This club sold twenty-five-cent tickets to give to the hostess for a dance. After a few too many beers, the sergeant returned from the restroom and said, "Look at my shoes and see if you see anything strange."

Even in my condition I could see he had put on his prosthesis backward, so the toe of his right shoe was pointed behind him. He smiled and said, "Watch this, Billy."

He walked up to a lovely lass and asked her to dance. She didn't notice his shoes, but halfway through the song he backed away and asked, "Do you see anything strange about me?"

She looked down at those two shoes, let out a scream, and ran from the dance floor.

We didn't get any more attention from the ladies—my friend's prank took care of that—but the bouncer took an immediate interest in us. He requested we depart soonest.

I admired these men who had lost a limb but retained a sense of humor, mostly because I thought an amputation would spell the end for me. Unfortunately, the prospect of amputating my right foot was raised by the chief orthopedic surgeon at Walter Reed, Dr. Arthur Metz. After three months, the damned thing still hadn't healed, and amputation would surely end my military career forever.

I knew I had to get busy to lie, cheat, and cajole my way out of this mess.

So busy I got.

*   *   *

My first order of business, strangely enough, was to ask for a ninety-day convalescent leave, to begin the latter part of July 1965. Dr. Metz accommodated my request, recommending I take my bottle of penicillin and head for my Texas home for three months of solid and uninterrupted rest.

I had a different plan. While receiving my convalescent-leave orders, I had the presence of mind to include several countries listed in the "allowed to visit" section. Among those included was the Republic of Vietnam.

The day I began my leave, with my leg and ankle still infected and draining pus, I began a four-flight journey that ended at Tan Son Nhut Air Base, approximately ten kilometers from Saigon. Once back in my element, I immediately sought out my old SF friends for some conversation and good times.

The place to be in Saigon was Tu Do Street. Only four blocks long, there were dozens of bars, and The Sporting Bar was the favorite of my SF friends. The Sporting Bar provided U.S. military personnel refreshments and a woman's company, twenty-four hours a day.

One night in August, as my friends offered many toasts to my ability to hop into South Vietnam, we lost track of the midnight curfew for U.S. military personnel in Saigon. Around 0030, two uniformed U.S. MPs, one a white sergeant, the other a black corporal, entered The Sporting Bar and asked for us to provide late-curfew authorizations.

My buddies were assigned to a new unit called
Studies and Observation Group (SOG), a top-
secret outfit run by the CIA. SOG was activated
in 1963–64 with the goal of undertaking guerrilla warfare in North Vietnam without the use
of U.S. personnel on the ground. My friends
were training NVA turncoats or South Vietnamese soldiers who supported our cause, and one of
the perks of SOG was a "Walk on Water"
(WOW) pass. These lads flashed their WOW pass
to the sergeant and they were home-free. The
WOW pass—with its twenty-four-hour-a-day
telephone number to address any questions—
was an open invitation to the entire war, author-
izing the bearer to be out at any time in any area
of South Vietnam. It allowed him to carry
firearms. It allowed him to order any policeman
or official to transport him to any location. Unfortunately for me, I didn't have a WOW pass or
a legitimate reason for being in the country or
The Sporting Bar.

So when the MP corporal, doing his duty,
asked for my ID, I promptly reached into a
pocket and produced my convalescent-leave orders, pointing out the provision that authorized
unlimited stay in South Vietnam and other nations. He looked at the location of my orders—
Walter Reed General Hospital, Washington,
D.C.—and then he looked at me. I could practically see the wheels of his brain coming to an immediate halt.

I tried to help. "I'm assigned to Walter Reed

General Hospital, in the Washington, D.C., area."

He looked at me again, with the same perplexed expression.

Without addressing me, the MP corporal walked over to the MP sergeant. He lowered his voice, and under his breath he said, "This motherfucker has these funny-looking papers from some fucking hospital in Washington. What do you want me to do?"

My friends were enjoying the drama, and by this time so was I. After all, what were they going to do with me? Send me home?

"What in the hell are you doing over here on a convalescent leave, Master Sergeant Waugh?" the sergeant asked.

"Sergeant, I came to visit my old Special Forces pals."

He thought hard for a few seconds and called for his radio. The sergeant spoke to someone named "Provost 6"—the commander of the military police, since "6" designates the commander of any unit. After a disbelieving pause, Provost 6 said, "Stand by."

We sat there for five minutes before Provost 6 came back on the radio and asked, "Is the individual in question causing any problems?"

"No sir, no problems, just out after curfew," the sergeant said.

"Well, then," Provost 6 said, "if he's crazy enough to be in Vietnam on convalescent leave, and his leave orders are good, leave him alone."

The sergeant signed off, refolded my leave orders, and gave us a hand salute before departing. As they left the bar, the corporal continued to shake his head and mumble under his breath. He didn't understand me or my motivation. He wasn't the first. He wouldn't be the last.

After a couple of weeks in Vietnam, I followed Dr. Metz's orders and returned to Austin, Texas, in September 1965 to conclude my convalescent leave. The cast on my right leg was practically destroyed, and the infected foot was draining through the cast. The smell, of course, was incredibly foul. I spent four or five days at my sister's home in Bastrop, then hopped a flight to Fort Bragg, North Carolina, to have the cast removed and replaced before I reported back to Dr. Metz and Walter Reed. With a new cast in place, I reported to Walter Reed a few days early.

The trip to Vietnam rekindled my efforts to return to the war zone as something more than a spectator. On a trip to the Pentagon, I met Billie Alexander, a great lady who served as the assignments officer for enlisted NCO personnel of the U.S. Army Special Forces. This meant she had control over every Green Beret, which made her position of high importance to me. The fact that Billie A. and I became awfully close in a nonprofessional sense didn't hurt my cause, either.

There was a new section of SOG in the planning stages, according to Billie, and this one would stand a better chance of success than the abject failures that preceded it. This new unit would be top secret, but I discovered SOG would be initiating Operation 35, an American-led outfit that would conduct missions behind enemy lines in Cambodia, Laos, and North Vietnam, rescuing downed pilots and working against the NVA along the Ho Chi Minh Trail.

It didn't get any better than this. I hadn't felt this sure of anything since I joined Special Forces in Germany in 1953. Wounds or no wounds, infection or no infection, limp or no limp, I simply *had* to be in this unit. There was no doubt about it. I had top-secret clearance, and my particular specialty—Special Forces master sergeant, intelligence, and weapons—was in high demand for this operation. In fact, it looked like they put together Operation 35 with me in mind. This was combat action at its absolute best, and I would not be denied entry.

Billie A. needed 120 senior SF NCOs for this new arm of SOG. I begged her to put me on orders, to which she replied, "Goddamnit, Billy, you're limping around now. Your foot and leg aren't healed. There's no way!"

I wouldn't give up. I cried, cajoled, and goaded, then I begged some more. I proposed a deal: She would reassign me to duty in the 7th Special Forces Group (A) at Fort Bragg. After I pulled duty there for one month, she would send orders reassigning me to SOG, allowing me to get over there in time for the new ground operation.

"OK," Billie said. "If you can pull duty at Bragg for a month without ending up in the hospital, I'll do it."

The next step was to get my sorry ass released from Walter Reed. I had summarily rejected a medical discharge, and just one week after I worked my magic with Billie A., Dr. Metz came to me with bad news. The infection wasn't responding. The foot had to be amputated.

He was serious this time. I convinced him to give me one more chance. I asked him to remove the cast from my right leg and allow me to continue with the penicillin until I was certain I was pure as driven snow. Amazingly, the foot responded. After a couple weeks, I found I could limp without crutches. As the time grew near for a final decision on the removal of the foot, just above the ankle, I grew ecstatic when the draining of the wound slowed and then ultimately ceased. The wound closed with about thirty pieces of shrapnel in the foot and ankle.

So in December 1965, I was reassigned to the 7th SFG (A) in Fort Bragg. I reported to duty with a definite and noticeable limp, but within a week I had orders for SOG. This was the news I'd been waiting to hear ever since that med-evac bird lifted me off the ground in Bong Son. I kept my body and all its parts intact and conned and connived my way back to the war zone. I headed for Vietnam one happy soldier. I traveled with a lot of unauthorized metal in my foot and leg, but none of that mattered.

I was hardheaded, dedicated, and persistent, and I used all three qualities to get my way. Damned if I wasn't going back to Vietnam, a little wobbly in the legs but determined in the mind. And this time, I wasn't going back to visit. I was going back to combat.

When I reported to the SOG commander of Command and Control North (CCN) in Khe Sanh during May of 1966, I couldn't hide my lack of mobility. I was limping, and no amount of machismo could disguise it. The commanding officer immediately sent me to the SOG training site southwest of Saigon in the town of Long Thanh. My assignment—airborne coordinator of SOG operations—did not appeal to me in the least. It was out of the States and back in the war zone, which was wonderful and against all odds, but beyond that there wasn't much to recommend the job.

I was in the right place, but the work—training and testing airborne personnel and equipment—was not where the bullets fly. I needed to be with the Special Forces men, in the thick of it. With so many great opportunities out there in the teeming, breathing jungle, I needed to be part of it. I was back on the team, but I needed to return to the field.

Euphemistically and insipidly titled, Studies and Observation Group oversaw the top-secret, dangerous, and demanding missions behind enemy lines. SOG conducted countless recon and rescue missions along the Ho Chi Minh Trail in North Vietnam, Laos, and Cambodia. At times the feats performed by SOG team members in rescuing personnel behind enemy lines bordered on the superhuman.

A brief and incomplete history of SOG: In 1965, the decision makers at the White House understood that Ho Chi Minh and his commanding general, Vo Nguyen Giap, were preparing to move men and materials from the north into the south. They would move them on the interconnected and labyrinthine passes, trails, and roads that ran down the western border of North Vietnam, into Laos. They were covertly constructing jungle-covered roads to accommodate trucks, tanks, people, and gear. Equipment could be man-packed or carried by bicycle into the south of Vietnam, all in preparation for an eventual takeover. Of course, this grid of roads and trails became known as the Ho Chi Minh Trail.

Our U-2 spy planes and C-130 high fliers watched these movements during nighttime reconnaissance flights. The U-2s provided photos of the HCM Trail being constructed. Through the Military Assistance Command Vietnam (MACV) and the CIA, the bosses of SOG recommended that small reconnaissance teams, consisting of three or four U.S. Army Special Forces and five to six indigenous personnel, be infiltrated into the border areas of Laos to watch the roads and trails. It was suggested they be inserted at various locations from the southern border of Cambodia or the Laotian/Vietnamese border. Thus Operation 35, headquartered in Saigon, was born.

SOG's mission was to counter Viet Cong and NVA guerrilla tactics with some of our own. We were given the opportunity to operate slightly outside the normal chain of command, to conduct our missions without concern for the day-to-day bureaucracy of the U.S. military. Funded by the CIA, SOG took its orders directly from the very top man in the U.S. government. I wasn't a politician, I was a soldier. Did I want to be a part of such an outfit? Damn right I did, and that meant putting aside my physical ailments and working my way back to the field of combat. My wounds, however, were once again a problem. Not only was I limping, but the right foot and ankle developed another infection and the draining began once again. I tried like hell to cover it up, and to a large degree, my concealment tactics worked. And once it started to feel better, I decided to bide my time until I could work my way north, to the SOG ground war.

Two and a half months was all I could take before I made a trip to the office of Colonel Arthur D. (Bull) Simons, the commanding officer of SOG. Simons was a combat-tested army Ranger in World War II and the commander of the top-secret "White Star" CIA mission in Laos from 1959 to 1961. He gained fame in November of 1970 for his work in organizing and commanding the Special Forces' Son Tay Raid, a daring rescue attempt at a prisoner-of-war camp near Hanoi. Simons gained even greater fame after retirement, when he led a 1979 rescue mission to free two of Ross Perot's employees who had been taken prisoner by the Ayatollah Khomeini's Iranian government.

You didn't trifle with Bull Simons. He was

large and forceful, with a resounding voice and a commanding air. He did not suffer fools, and if you gained a meeting with him, you'd better have your story straight and ready. He looked at me with suspicion when I requested he assign me to a role more befitting my abilities. I awaited the expected onslaught from the Bull, and I wasn't disappointed. His suspicion turned to scorn.

"Billy Waugh, you dumb shit," he said by way of salutation. "You are still limping, and what the hell do you expect me to do about that?"

I didn't reply, and my silence seemed to settle him down some. He sighed and thought about my request. I couldn't tell exactly what the response would be, but I grew more optimistic the longer he took to formulate his reply. I guessed if the Bull was going to tell me to go to hell and throw me out of his office, he wouldn't need much time to think about it. I knew I needed to fire off a few quick words describing my abilities and competence, so I did this and finished by blurting out, "I will do you a great job, Colonel."

Finally, he said, "All right. Get your ass up to Khe Sanh. Take over the launch site as coordinator, until you're completely healed. Any fucking questions?"

Simons's call sign was "Dynamite," so I responded, "Roger that, Dynamite."

"And one other thing, Waugh," he said. "I don't want to hear 'Mustang'"—my SOG call sign—"over the radio whining to come down here to Saigon to get drunk. As a matter of fact, I don't want to hear a fucking peep out of 'Mustang' for six fucking months. Now get the fuck out of my office."

I rogered that transmission, too, and left the office one happy man. It was July of 1966, a little more than one year after being left for dead on a rice paddy in Bong Son. I departed Simons's office with a little bounce in my battered step and went to pack my bags. I was off the bench and back on the playing field.

At approximately 0900 on August 4, 1966, I looked down from the back seat of a SOG Cessna to see the body of Sergeant Donald Sain staked and tied to a bamboo tree, arranged in a spread-eagle position on the ground in a clearing in Laos's Co Roc mountain area. The NVA bastards had killed Sain and Staff Sergeant Delmer Laws, both members of Recon Team Montana, and they decided to send a message by putting Sain's body on display, like a showpiece.

This is how my first rescue mission as a SOG man started: with anger. I had been in combat long enough to know Sain's body was not only displayed so we would see it, but it had to be booby-trapped as well. The NVA wanted us to find this body, and they wanted us to blow ourselves up trying to retrieve it. Sain and Laws were the first SOG recon men to be killed in combat with the enemy, imbuing their deaths with added significance. It pissed me off that the NVA would use one of our men as a booby trap, and it pissed me off even more that they thought we were stupid enough to fall for it.

Recon Team Montana—the first SOG recon team to launch from Khe Sanh and one of the first recon teams in the unit's history—infiltrated by helicopter into Co Roc at last light on July 26. The nine-man team was led by Master Sergeant Harry Whalen, an experienced and hardened combat veteran. His team consisted of Sain, Sergeant Delmer Laws, and six indigenous soldiers. Montana spent eight mostly uneventful days on the ground in the Co Roc region, roughly fourteen kilometers from the Khe Sanh SOG camp. The team watched and reported on NVA activity in the area. Their orders were to gather intelligence and avoid initiating combat. Montana was to fire only when forced.

Each team member carried a minimum of twenty-five magazines of 5.56 mm ammunition for their submachine guns, several grenades or minigrenades, sawed-off shotguns, sawed-off 40 mm grenade launchers, as many pistols as they could handle, M-14 mines (called toe poppers), and any other lethal device they could carry. The SOG recon teams were loaded for bear, and the NVA knew that closing with the team in combat amounted to suicide.

The principles of SOG hinged on one fact: The jungle is both friend and foe to the small unit that moves and lives within it. The jungle is a life force, more than just landscape. It must be respected for its power and exploited for its mysteriousness. The jungle hides enemy structures, roads, and personnel, but it also allows a friendly team to infiltrate and become lost in the vast, dark cover. SOG created soldiers who would embrace the cloak of the jungle.

For eight days, from its infiltration on July 26 through August 3, Montana reported no unusual incidents. But in the jungle, fortunes can change quickly and without warning. At some point on the afternoon of August 2, 1966, the nine Montana team members were ambushed by a platoon of NVA estimated at thirty-five soldiers. Montana walked in a well-spaced diamond position, and the NVA opened up on the nine men with automatic weapons fire that ripped through the jungle's vines and trees. The NVA was too well organized for Montana to counter the ambush, so Whalen and six indigenous shot their way to a rallying point nearly two kilometers from the ambush site. There they waited until they could sneak, scurry, and scramble down the east side of Co Roc Mountain, out of Laos, and back into South Vietnam.

Montana's failure to make proper radio contact on August 2 alerted us that the team was in trouble. The next morning, Whalen and the six Montagnards (the mountain people who lived in Vietnam's central highlands) straggled into the Khe Sanh Special Forces camp. Ragged and tired, Whalen was immediately taken away to a restricted SOG site for debriefing. Sain and Laws were nowhere to be found, and Whalen reported that he suspected they had been killed or captured. Whalen provided the approximate location of the ambush, and a search and rescue began immediately. While Whalen was being debriefed, a pilot and I left the Khe Sanh airstrip in the USAF Cessna 0-1, searching for Sain and Laws. Within minutes, we were over the mountainous terrain, looking over the right side of the plane at the NVA's calling card: Sain's exposed and dead body.

There's a code in Special Forces that I need to make clear: If you die, we're going to get you out. We do not leave wounded or dead men behind for the enemy to gloat over, stake out, or other such bullshit. We leave no man behind. It's as simple as that. If it costs the lives of fifty men in an attempt to get you out, we're still going to send the 51st in there to retrieve all or part of your body.

This code began with the advent of Special Forces in 1952 and exists to this day. This policy was strained during Vietnam, since the terrain and the remote nature of much of our SOG warfare left bodies in unreachable places. This was all new, and we had to improvise. In infantry days, a body that fell amid a company was much easier to remove from the battlefield. It was often difficult to get a helicopter into some of the places where SOG men went down, and if it wasn't for the heroism of many of our Vietnamese Air Force pilots, and those rescue-and-recovery teams within the birds, many more bodies would have gone unrecovered.

When we landed at the SOG site in Khe Sanh, we briefed Major Jerry Kilburn on the situation

SOG TARGET AREAS (LAOS)
    GENERAL MAP OF THE TARGETS OPERATED AGAINST
    FROM THE LAUNCH SITE AT KHE SANH, SVN

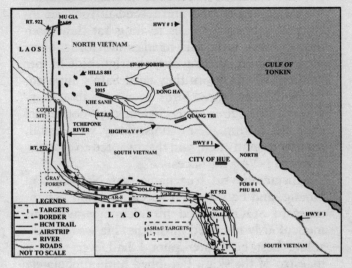

regarding Sain's body, which we knew was not only displayed prominently but certainly booby-trapped. Laws's body was nowhere to be seen.

Kilburn put me in charge of the rescue. Staff Sergeant Danny Horton and Sergeant First Class James Craig were placed under my command. Kilburn would also accompany the rescue team. The lead H-34 pilot of the Vietnamese Air Force, a supremely talented pilot nicknamed Musta-chio, was the unanimous choice to fly into the re-covery zone.

These VNAF pilots were remarkably talented. They were the cowboys of the air, able to make their H-34 birds perform magic in the skies and on the deck. These men were former jet pilots who were trained in Texas to fly the SOG H-34s for jungle infiltration and rescue. It took an incredible amount of balls to do what these men did, and they were paid handsomely for it. Their deal called for them to be paid by the CIA per landing, which meant they needed very little enticement to put those birds down just about anywhere. Every time they landed, they could feel the cash hitting their palms. They were good, they were motivated, and they wanted to stop the detested NVA's advances.

Mustachio—his true name was Nguyen Van Hoang, and he was on a special-mission assignment to SOG—earned his nickname with his meticulously trimmed mustache. He was the absolute best helicopter pilot I had ever seen, a maestro of the skies. I grabbed him immediately for this rescue operation, because I knew this was not going to be a conventional body recovery. The extraction of Sain's body—located in a small clearing not large enough to put the bird on the ground—would take a pilot with great dexterity, skill, and courage, and my friend Mustachio possessed all.

We arrived over the vicinity of the body and searched for a landing zone. The five of us looked down from the H-34 at Sain's body as we circled the small clearing where his body was tied and staked. I could feel the anger and disgust as it exuded from each man. Kilburn and I decided to land approximately a hundred meters from Sain's body, and after a quick touchdown, the rescue team hopped out of the helicopter and moved toward the body. I moved quickly and forcefully, the pain from the shrapnel in my legs erased by the adrenaline of the moment.

We reached Sain's body and approached carefully. Several hand grenades, with pins pulled, were placed under his body, ready to detonate if we moved the body in any manner other than vertical. Sain had been dead in the jungle heat for more than thirty-six hours. He died of gunshot wounds to the chest, and maggots had filled his wounds like putty. Flies were buzzing in and out of his nose. The maggots were all over his bloated and decomposing body. The stench of rotting flesh was nearly unbearable.

I grabbed a climbing rope and told Kilburn, "Here's what we're going to do. I'm going to tie this around his leg and tie it to the wheel of the H-34. We'll move the body to the LZ and take it from there."

Kilburn and the rest of the team looked at me like I was crazy. To them, it sounded like a wild stunt that would never work. These guys hadn't seen the combat I'd seen, and they weren't accustomed to thinking on the fly, to doing what needed to be done no matter how unusual it might be. There's no manual for this kind of horrid shit. But knowing Mustachio's prowess, I felt this was the fastest and most efficient way to achieve our objectives.

I explained the situation to Mustachio and tied the rope around Sain's leg. I ordered Mustachio to hover over the body as low as possible.

He lowered the bird over the body like a man palming a basketball, and I tied the rope to the wheel of the H-34 while he continued to hover.

"OK, let's move out," I told the team members. Mustachio then lifted the body off the ground while two or three of the booby traps exploded harmlessly. Sain was hovering above us by his leg, and all the fluids and maggots and shit from his body flowed out as he was being lifted off. Kilburn, who apparently had forgotten some of the more unsavory facts of the human body, screamed as his face and body were showered with the detritus of death.

Mustachio lifted Sain's body out of the combat area and set his bird down in an open space away from the jungle. With Sain recovered, we began the hunt for Sergeant Laws. I ordered the team to fan out, keeping the man next to them in sight, and search for the body. We fought our way through the jungle, tearing up our arms and faces on "wait-a-minute" bushes—bushes with sharp thorns that sliced through the sleeves and legs of our fatigues like scythes when we slashed through the jungle. To get past them, you had to wait a minute while the man in front fought his way through the thicket. If you didn't keep three paces behind, thorns slashed across your forehead. The damned jungle was a son of a bitch.

Monkeys in the trees and bushes jibberjabbered at us, their singular focus on our activities a good sign that no NVA platoon was in the vicinity. Mustachio continued to hover away from the area, but within radio range, prepared to move to the landing zone on my signal. We searched and searched, finding trails and massive blood spots along the way. We searched for hours, so long that Mustachio left Sain's body in the clearing and returned to Khe Sanh to refuel. We stayed on the hunt, though, and finally, after several hours, Danny Horton called to me from out of the bush.

In seconds he emerged from the cover holding Laws's left leg by the jungle boot.

"Billy," he said quietly, a sick look on his face. "This looks like an American jungle boot."

Tigers prowled this area of Laos, and it appeared they found Laws before we did. I placed the leg in a body bag, and we continued our 360-degree search of the ambush site. I came upon several more massive bloodstains and numerous trails, but no NVA or bodies were found, and neither was any other portion of Laws's body. At approximately 5 P.M.—five hours into the rescue mission—the signal was given for the H-34 to exfiltrate the rescue team, along with Sain's body and Laws's leg.

We quickly covered the hundred meters back to the LZ. We set up our perimeter of defense, as the H-34 approached with Sain's body. I untied the body from the rope as the helicopter touched down. Sain's bloated and misshapen body was placed in a body bag and into the H-34. The left leg of Delmer Laws was placed in a body bag as well, and the rescue team climbed aboard and returned to the Khe Sanh Special Forces camp. The ride was quiet.

*   *   *

Montana was one of the first SOG recons team
to spend considerable time behind enemy lines in
Laos. Sain and Laws will always be remembered
as the first of many fine American men to lose
their lives in the battle for the Ho Chi Minh
Trail. We didn't know it at the time, in the quiet
contemplation of the trip back to Khe Sanh, but
this rescue would be the first of many.

When we disembarked in Khe Sanh, the rescue
team was a mess. I stunk like shit from the de-
composed bodies. The stench of death hung on
us like another layer of clothing. Kilburn was the
worst, of course—the fluids and solids from
Sain's body had showered him, and the smell was
enough to make anyone within fifty feet retch.
The sleeves of my fatigues were shredded like
confetti from the wait-a-minute thorns, and my
face and arms were scratched like I'd been in a
losing battle with an angry cat. I looked horrible
and felt worse, having seen one of our fine men
booby-trapped and being used as a showpiece by
the NVA.

The command sergeant major of the 5th Special Forces (A) was at Khe Sanh, taking a break
from his desk duties at HQ in Nha Trang to visit
an SF A team that shared the base. This man was
a notorious starched shirt, changing his fatigues
as often as three times a day to ensure the proper
look. He was a stickler for details and decorum,
and even though he had no jurisdiction or interest in SOG operations, he saw my condition after
getting off the rescue chopper and couldn't help

himself. He walked up to me and gave me a condescending once-over.

"Goddamnit, Waugh, you're filthy and you're out of uniform," he said to me. "I've never seen anything so disgusting in all my life. You guys are horrible-looking excuses for soldiers."

I clenched, trying to contain my anger. We had just recovered two bodies, and we had to hang Sain up in the air for more than two hours in order to do it. We recovered a leg attached to a U.S. jungle boot with the full knowledge that animals had eaten the rest. Kilburn could still taste the shit and juice in his mouth hours later.

Finally, I said, "Look, you got your starched shirt on. You work down in Nha Trang, behind a desk. Goddamnit, I work out in the fucking field. That's the difference between you and me."

Then I reached down and unzipped the body bag that held Laws's leg. I pulled it out of the bag and held it by the jungle boot, right in front of him. I let it hang there.

"How about this guy, Sergeant Major?" I asked him through tight lips. "Is he out of uniform? This leg is all that is left of one of our SF men, for the rest of him is in the fucking jungle where we just came from."

I held that leg as firmly as I held my stare, then lowered my voice and said, "How about leaving me and my team the fuck alone, Sergeant Major?"

I thought he was going to throw up. Finally, he said, "I'm sorry, Billy. I shouldn't have said that."

"Goddamn right," I said.

"Billy, what can I do for you guys out here? What do you need?"

"Well, we need some uniforms," I told him. "These uniforms are awful. That would be a start."

Within days, we received a package through Special Forces channels: three boxes of brand-new jungle fatigue uniforms.

Whalen, a World War II veteran and one of the most cynical humans I have ever known, took the loss of Sain and Laws especially hard. He was a team leader who returned without two of his men, which can be unavoidable but no less traumatic. Whalen spent many nights drinking heavily in a futile attempt to shed the memory. That first night, in fact, a group of recon men drank with him, toasting their fallen friends.

In that bar a SOG tradition began. The recon men began singing a song called "Hey, Blue."

I had a dog and his name was Blue
Bet you five dollars
He's a good dog too
Hey, Blue
You're a good dog, you

From there the song took on a life of its own. Blue's loyalty and friendship is recalled, and Blue dies in the final verse. In the final verse, the men substituted the names of Sain and Laws:

Sain and Laws
Hey, friends
You were good guys, you

This song became a regular at the SOG wakes, with more and more names of SOG KIA added to each singing. I wasn't there for the impromptu memorial for Sain and Laws, and I don't recall being there for anybody else's, either. I had to keep working. I was the man who led the effort to rescue their bodies, but I couldn't bring myself to celebrate their lives. I had seen too much death. I was the guy who went from one job to the next, the guy whose single-minded pursuit overtook everything else, the guy who never took the time to either remember or forget.

Over the years many of my troops looked me in the eye and said, "Sergeant Major, with all due respect, you are one cold fish." My calculating actions in combat led the men to believe that I was without feelings or remorse. For instance, I was not delicate with Sain, for Sain was dead. I chose to lift his body out with the climbing rope rather than by hand for a simple reason: Dragging the body out by hand would cost us if the enemy opened up on our rescue team. I was not a cold, calculating fish, but I was a man who was always busy thinking of the next move. While others mourned, I studied ways to keep men alive, how to complete the mission, how to solve the problems that presented themselves every goddamn day in the unforgiving jungle. I am a problem solver, and sometimes problem solving must take place at the exclusion of sentiment.

So I mourned the men in my own way. By the time I joined SOG, I had already lost men in combat under my command. Ridding my mind of the ghosts of KIA was never a reason for lifting a beer to my lips. I enjoyed the carefree ambience of drinking during nonwork time, probably because I was so utterly consumed by the work. I was never much for the camaraderie of death.

CHAPTER 3

I am not a man who sits back and ponders the effects of war on the human psyche. I am a man of action, a man who *functions,* and during war I never spent time gazing into the distance, pondering the meaning of it all. I have seen men made and unmade on the field of battle, and I understand its power. I just never saw much sense in bemoaning my fate or blaming someone else for the pain I feel from the shrapnel in my legs. I was a soldier. I had a job to do.

During my seven and a half years in Vietnam, I did not concern myself with politics. We were aware of the sentiments at home, and we saw firsthand how the protests and general unrest boosted the enemy. But in a very literal sense, this was not of our world. The way I saw it, this was the only good war at the time, and by damn I was going to be there to fight it. I understood the power of war, and as such I understood that a soldier can never, ever, be complacent in searches across the combat area. Just when you think you have the entire world by the ass . . . bam bam, you are dead. In that world, the world of the SOG men, there is no room for politics. So I did not question why we were in this war. Instead, I channeled my energy toward trying to win it.

And June 1, 1967, was a day that held the promise of something momentous, something that could alter the entire course of the war. On that day, it was evident something important was happening in a small Laotian valley dubbed Oscar-8. The National Security Agency reported fifteen hundred special agent reports (SPARs) emanating from this spot about twenty kilometers southwest of Khe Sanh. U-2 high-altitude recon flights reported a huge increase in vehicular traffic in the area during the hours of darkness. The NSA identified Oscar-8 as a high command field headquarters for the NVA, and the rise in radio transmissions intended for Hanoi high command led SOG to believe NVA commander Vo Nguyen Giap was paying a visit to Oscar-8.

Giap was the key to the war. He had a sharp mind for combat, and I grudgingly admired his savvy approach to achieving his objectives. He was the brains of the outfit, and I—as well as those above me in SOG—believed his death would have changed the course of the war. Cut off the head and the body will follow, as it wiggles to certain death.

The prospect of Giap's presence at Oscar-8 prompted SOG headquarters to order Command

and Control in Da Nang to strike this target. The strike would start at 0600 on the appointed day with nine B-52 bombers dropping approximately nine hundred 200 kilogram high-explosive bombs on the target. Two SOG Hatchet Forces (fifty-five Montagnards and four U.S. Special Forces) would follow, infiltrating the Oscar-8 area via Marine CH-46 helicopters. The objective was bold and decisive: Kill Giap and all other enemy forces encountered along the way.

It would not be easy. Oscar-8 was a defender's dream, a bowl-shaped valley surrounded on three sides—north, east, and west—by hills that formed a horseshoe. The hills ranged from roughly eight hundred feet to more than fourteen hundred feet above the valley floor. The valley was roughly one kilometer wide and three kilometers long. Giap was a crafty old bastard who had been in combat most of his adult life, and he undoubtedly saw the protection Oscar-8 afforded. Route 922, a dirt road to the south of Laos, ran through it, toward Vietnam's notorious Ashau Valley. The floor of the bowl was relatively flat, with heavy jungle canopy. Place antiaircraft weapons on the high ground and the valley floor below became nearly impenetrable to an invading force. It was a dream encampment for a clever defender with a lot to defend.

All of our tactical air resources were in place—HU-1D marine helicopter gunships to clear LZs, A-1E Skyraider propeller aircraft to support the Hatchet Forces, four F-4C Phantom jets for close air support, two SOG H-34 search-

and-rescue choppers, two 0-2 forward-observer planes for support. The Hatchet Forces would arrive at 0700 on June 4, 1967, and sweep the area of Oscar-8 following the devastation caused by the B-52s. They would capture or kill the commander, then withdraw no later than 1500 that same day. Eight hours of unbridled force would be unleashed on Giap and his NVA friends, ending with a Bomb Damage Assessment (BDA) that would confirm the death of the commanding general of the People's Army of Vietnam.

At 0400 on the morning of the attack, I lifted off in an 0-2 FAC aircraft piloted by U.S. Air Force major James Alexander. We were the first to survey the target area, and as we flew the thirty-five minutes from Khe Sanh across the Laotian border, I felt the adrenaline of combat well up inside me. I had been in Vietnam off and on for more than three years, and looking down at the all-encompassing green of the Southeast Asian jungle, I thought to myself, *This day could change the war*.

We approached the target in the 0-2 from the west, and we loitered approximately fifteen kilometers to the south of Oscar-8 to await the forthcoming B-52 strike. My interior anticipation was tremendous, and I couldn't wait for the show to start. At 0445, we looked into the early morning purple-orange dawn and saw contrails approaching Oscar-8 at about thirty thousand feet. The awakening sky, turning orange as sunrise neared, made the target vaguely visible as the contrails approached, and I spotted several early morning cooking fires in the Oscar-8 encampment.

We loitered airborne from a safe distance as about nine hundred dumb-as-dirt bombs were dropped from three sets of three B-52s at precisely 0600 hours. One after another they fell, deadly pellets from the sky. As the last bomb exploded, Alexander and I departed our loitering area and flew with speed and purpose to a spot above the target area. I looked down at the bomb craters and the numerous fires burning below. Oscar-8 felt the wrath of our weapons, and I felt like God, watching with a divine detachment.

Dozens of NVA scurried to put out fires that burned from Route 922 to the base of the high ground to the northwest. Due north there were fifteen to twenty men attempting to roll fuel barrels away from a ferocious fire in the NVA fuel-storage facility. Several grass shacks and houses were burning, and people in every manner of dress and undress were scattering like roaches suddenly exposed to light. Alexander and I hadn't worked together before, but the enormity of what we were witnessing forced an immediate bond.

The NVA quickly put itself in position to defend its prize encampment. Quicker than we thought possible, an onslaught of withering 12.7 mm automatic antiaircraft weapons fire began to sear its way past us on every side. Given our position over the bowl-shaped target area and the altitude of the surrounding hills, the antiaircraft fire was actually being fired *down* at us. Major Alexander immediately showed me his aptitude by juking and dancing that bird around the ceaseless attack.

This speed and aggression of the NVA's response startled me, and my thoughts immediately focused on the wisdom of dropping the Hatchet Forces via helicopter into such a heavily guarded area. The B-52 bombing had done significant damage, but it clearly had not destroyed the NVA defenses. The CH-46 troop-carrying helicopters were scheduled to land in random LZs in Oscar-8 within fifteen minutes of the last B-52 bombing, and as the antiaircraft fire flew on each side of us, I grabbed the radio and frantically attempted to contact the Marine pilots manning the CH-46s.

I looked to the west and saw two Marine gunships strafing an area intended to be used as a landing zone for the CH-46s. We watched helplessly as ground fire picked off both of the gunships and blew them from the sky, side by side. The gunships wobbled, one after the other, then crashed on the landing zone like a choreographed nightmare. This sight of these gunships going down hit us like a direct shot of adrenaline. Major Alexander and I called on UHF and VHF radios to abort the landing of the troops. We were frantic, screaming. Those Hatchet Forces could not land. It was suicide.

"Abort! Abort!" I screamed into the radio. "Do not land! Abort those fucking choppers!"

Goddamnit. They couldn't hear us. The Marine pilots must have switched radio frequencies. Our words went nowhere. It was too late. We were helpless. A buzz ran down my spine. The hair on my arms stood up. The landings were in progress, and the Hatchet Forces were heading blindly and unknowingly toward a fucking bee-hive of NVA activity.

Alexander and I continued to scream point-lessly into the radios as two CH-46s were shot down while airborne, approximately fifty to one hundred feet in the air. *Fuck!* Both were shot into two separate pieces. Three aircraft down. The NVA guns continued to shoot down at the targets picking them off with ease. *Son of a bitch!* I was helpless, sitting there with a fucking worthless radio, watching our men go down.

I dropped the radio in my lap as I watched our troops tumble out of both aircraft. They fell onto the landing zone like rocks.

Alexander rose to four thousand feet and to the west, out of reach of NVA antiaircraft fire, to make clear contact with Hillsboro. I reached the commander of Command and Control, who was at the site in Khe Sanh.

"We've got a real problem out here at Oscar-8," I said.

I continued to scan the Oscar-8 horseshoe area, my head on a swivel, my eyes darting and huge with disbelief. This was the three-ring cir-cus, and I fought to keep up.

NVA fire from the ground and the hilltops had our aircraft caught in a deadly crossfire. Enemy weapons took on every bird in the sky. Two H-34s attempted to move into the LZs to pick up survivors from the downed Marine helicopters. One was raked with enemy fire. It burst into flames immediately and crashed directly onto Route 922. Its three crew members scrambled from the burning bird and moved up the hill that rose to the south of Route 922. I spotted them and immediately passed their location along to the net control in Khe Sanh.

We were the eyes of this operation, watching in horror as the plan fell apart below us. We directed two F-4C Phantoms over the rims of the horseshoe ridge. One Phantom covered the second as it struck the target area. I couldn't believe my eyes as I watched, dumbfounded, as the Phantom was struck in the right wing with AA fire. The wing exploded, going up like a matchstick, and the plane followed suit.

No parachute appeared. The proud bird crashed to the ground, reduced to nothing more than twisted, random pieces of metal.

Two A-1E Skyraiders were next in line. They were providing air strikes along the horseshoe ridge when one of them, passing low to drop napalm directly on the antiaircraft guns, was chopped apart by the very guns it was attempting to destroy.

Again, no parachute was visible as the aircraft crashed into the ridgeline.

I received a call from a member of a Hatchet Force on the ground. He pleaded for tactical air. I saw the team's position, marked by two bright red panels dropped at the edge of two bomb craters that sat about fifty meters apart in the center of Oscar-8. The team leader said he and approximately twenty-five indigenous team members were in those craters, craters formed by our own B-52s. The team leader said his team was holding its own against NVA attacks. He begged for tactical air to drop napalm and cluster bomb units within thirty meters of his position.

"I'll get it to you," I told him. "You guys hang in there."

I looked at my watch: 0900. It had been three hours since the initial B-52 formation blasted Oscar-8. So much had happened since then; it was like watching the world through a kaleidoscope. I once again called Hillsboro—the airborne communications post—and requested further tactical air for the area surrounding the bomb craters.

We were low on fuel. I asked Alexander to fly high on the return flight to pass information to Khe Sanh and Da Nang simultaneously. From there I reached Lieutenant Colonel Harold K. Rose, the commanding officer of SOG's Command and Control. His call sign was "Gunfighter," and he was in Khe Sanh awaiting word from the target area.

"Gunfighter, this is Mustang," I said. "Are you sitting down, Gunfighter?"

"No, Mustang. Why?"

"You need to sit down, Gunfighter, and you need to get out a pad of paper and a pen while I give you a count of aircraft lost."

TARGET OSCAR-8 U.S. AIRCRAFT SHOT DOWN IN THE TARGET

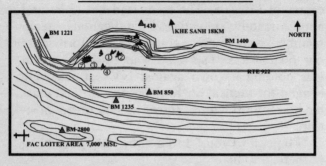

I recited the following: ① / ② Two CH-46s shot in two pieces (each) over the drop zone; ③ one HU-ID gunship destroyed on the LZ; ④ one H-34 shot down on attempt to rescue; ⑤ one F-4C phantom jet shot down with no parachute; ⑥ and one A-1E shot down over the target, with no parachute. ⑦ Depicts the bomb craters where the Hatchet Forces were defending their existence.

"Roger, Mustang," he said.

"Gunfighter, we've lost two CH-46s shot in two pieces over the drop zone, two HU-1D gunships shot down on the LZ, one H-34 shot down in an attempt to rescue, one F-4C Phantom shot down with no parachute visible, and one A-1E Skyraider shot down over the target, with no parachute."

I described the bomb craters where the Hatchet Forces were fighting for their lives, and the airwaves went silent for at least one minute. Finally, with a sigh apparent in his tone, Rose said, "What do you recommend, Mustang?"

"Gunfighter, we've got to bring up another Hatchet Force from Phu Bai, for reinforcement. Quickly call for no less than two SOG H-34 helicopters for rescue purposes. Hillsboro has ceased all bombing in North Vietnam and has at least eight sets of jets stacked up and orbiting within five minutes of the target. The horseshoe ridgeline is still being plastered with bombs."

"Roger that, Mustang," Rose said. "What's your status?"

"We are coming in to base for fuel."

"OK," Rose said. "Come to the bunker. I have my bosses on the line."

I "rogered" this transmission, but I have to admit I had something else in mind. My thoughts were occupied entirely by the crew of the shot-down SOG H-34, the three men I had seen scampering toward the jungle after the chopper crashed on Route 922.

A replacement FAC was in the air, and when Alexander and I landed at Khe Sanh the first thing I noticed were two H-34s, with blades turning, on the airstrip. I ran to one of them and jumped aboard, joining the three SOG Vietnamese crew members.

Mustachio was in the pilot's seat. Perfect. This was another mission that called for his expertise. I grabbed him by the leg and told him I knew where the H-34 crew was located south of Route 922. I had no commo, but I gave Mustachio a thumbs-up to get this bird off the ground to start the rescue mission. Without hesitation, my friend Mustachio lifted into the air.

Mustachio flew balls to the wall, first west, and then south toward the target area, while I directed him to the spot I had seen the three scurry out of the downed H-34. Once there, the Vietnamese crew and I scanned the area surrounding the burned-out chopper that lay in pieces in the middle of Route 922.

A handheld mirror signal flashed below us and to the south of Route 922. Mustachio cut power and nearly dived to the spot. Ground fire rose up to meet us, but we disregarded it as it whizzed harmlessly past us. We swooped into a small clearing, near the mirror site, and I saw the downed H-34's crew—three Viets—in an area surrounded by fernlike leaves and bamboo thickets.

There was an opening, barely an opening, roughly five square feet of opening, but that was enough for Mustachio. With the skill only he possessed, Mustachio lowered the bird down to the spot. The crew and I reached our hands out and one by one grabbed the hands of these downed crew members and pulled them into the H-34.

I gave Mustachio the thumbs-up, slapped his leg, and yelled, "Haul ass out of here!" He wasted no time. We lifted off in a hail of green tracers as Mustachio plowed through several small trees. Their branches burst apart like toothpicks as the rotor blades tore through them.

"Don't fly north!" I screamed at Mustachio, for a route north would take us directly over the horseshoe ridge of Oscar-8 and the deadly NVA antiaircraft firing positions.

I noticed the sound of the H-34 blades had taken on a different pitch. The consistent *thwack thwack thwack* of the blades was gone, replaced by a less consistent and more troublesome sound.

One or two of the four rotor blades was damaged by our trip through the trees, changing their bite. Nevertheless, we continued to limp through the air toward the west and south, gaining altitude as we gradually left Oscar-8 behind and headed toward the Khe Sanh airstrip.

I turned and smiled to this H-34 crew, and their relieved smiles in return said it all. These three men were damned happy. The dapper little command pilot of the downed H-34 was named Hiep, and he came over to me and put his arm around my neck. We looked directly at one another and said nothing. Nothing needed to be said.

I could imagine "Gunfighter" was having a frigging fit in the Khe Sanh SOG operation bunker, where I was supposed to be. The idea of his anger was tempered, however, by my knowledge of how pleased he would be at the rescue of these SOG helicopter pilots.

Through the confusion and mayhem of June 4, 1967, the enemy took a heavy pounding in Oscar-8. The nine-ship B-52 bombing had scored direct hits on the intended area. It destroyed NVA fuel dumps, buildings, and bunkers. The nine hundred HE dumb bombs wreaked havoc and disrupted the workings of the NVA headquarters area. The enemy's defenses and extraordinary protection in Oscar-8 convinced me that Giap had indeed been in the target area.

The remnants of the Hatchet Forces, trapped

and dangerous like rabid rats in the bomb craters, were a problem for the enemy. NVA commanders were aware of the small team's ability to direct tactical air strikes, and these commanders were not anxious to order a frontal attack of the craters. They surely would not attack before darkness.

I discussed the situation at the target area with Lieutenant Colonel Rose and Major Jerry Kilburn at the Khe Sanh bunker. At this point, at 1000 hours on the day of attack, fourteen U.S. pilots and crew members were on the ground. Radio reports indicated the team leader of the Hatchet Force was in need of a med-evac, and Rose decided Kilburn would be inserted into the target area to take over the leadership of that Hatchet Force.

Rose ordered four additional H-34 SOG helicopters repositioned from the Da Nang air base to Khe Sanh. He also ordered the two H-34s that had not been destroyed to insert Major Kilburn—along with a basic load of ammunition, grenades, and commo gear for the Hatchet Forces—into the bomb-crater area as soon as humanly possible.

As I headed back toward the 0-2 to resume my FAC duties, Rose caught up and put his arm around me. I stopped and we looked at each other, knowing so many of our fine men were hanging in the balance.

"Thanks, Billy," Rose said somberly. "Do the best to get the rest of them out."

"Roger that," I replied quietly.

At approximately 1600 hours, the insertion of Kilburn and the med-evac of the wounded team leader were completed. Eight seriously wounded Hatchet Force personnel were evacuated on the same bird that inserted Kilburn. This operation took place with the foreboding accompaniment of tactical air strikes on the antiaircraft positions along the horseshoe ridgeline.

Alexander and I returned to the air and searched fruitlessly for three hours. Our hopes were buoyed briefly when Hillsboro, the airborne command post, detected one emergency radio signal emitting from the area of the first rallying point. We spent forty minutes scouring the location, but in the vernacular of the U.S. Air Force, we received "No joy."

We flew over the target area, over Route 922, over the rallying points. We looked down at that vast world of green and strained to see something, anything, that would indicate U.S. military personnel, dead or alive. We searched and searched, our heads on a swivel, our eyes blurring with strain. Still, nothing.

The wounded Hatchet Force commander, upon returning to Khe Sanh, reported that he had seen SOG platoon leader Billy Ray Laney killed in action as the CH-46 crashed to the ground upon insertion. He also recounted seeing at least fifty NVA troops as friendly and enemy scurried about the battle area following the crash. The view from the ground, it appeared, was similar to the view I had from the air: utter chaos.

Night came quickly to Oscar-8. The valley
floor, surrounded by hills and covered by dense
jungle vegetation, was like a windowless base-
ment as late afternoon approached. Night was
not the friend of the Hatchet Forces as they de-
fended their position in the bomb craters. As
darkness fell, Kilburn requested a Specter support aircraft with its high-volume, high-speed
miniguns to provide ground fire and flare support through the night. Sets of F-4C Phantom
jets were prepared to deliver air strikes to the
horseshoe ridgeline beginning at nautical twi-
light and continuing throughout the hours of
darkness.

We were gradually learning more and more
about the situation within Oscar-8. Major Kilburn, now in a crater with the Hatchet Forces,
sent a situation report at 1700 hours, reporting
that two U.S. and twenty-two indigenous personnel were operating within the defensive
bomb-crater positions. Three of the indigenous
were slightly wounded but ambulatory. Ammo
was adequate and morale was good, considering.

With fourteen U.S. pilots and crew members
still on the ground, we couldn't pull the Hatchet
Forces out of the craters, so the decision was
made to use them as search-and-rescue patrols.
To put the situation in blunt terms, after SOG
got the Marine pilots and crew members into
this deepest of shit, SOG was not about to skip
out while those guys beat the bushes in an attempt to escape. Predictably, though, the effort
was unsuccessful. These Hatchet Force men,

pinned into the bomb craters by NVA mortar fire and snipers, couldn't leave those craters. It would have been suicide.

Air strikes prevented the NVA from attacking the craters. As one strike was completed, another rolled in with bombs, CBU, napalm, and 20 mm machine guns. Strike after strike after strike, with Kilburn calling in napalm within twenty yards of the southernmost bomb crater. The horseshoe ridge lines were continuously struck with every piece of ordnance in the USAF inventory. The effort to protect these men was as concerted as the effort to get them there in the first place.

Bombs dropped. Guns fired. Kilburn screamed into his radio, "Come on, Charlie, you fuckers! Here we are! Come get us!"

I continued to fly over the target area overnight. We served as the radio relay from the Hatchet Forces to the commanders at Khe Sanh and Da Nang. Everything else in the war had ground to a halt while this operation took center stage. Still, it was painfully obvious the operation needed to be aborted, and the Hatchet Forces needed to be rescued. The question, of course, was how. Lieutenant Colonel Rose and SOG headquarters ordered an ambitious and ballsy extraction plan, calling for four H-34s to land in the bomb craters, one at a time. They would rescue the Hatchet Forces personnel while another H-34 flew nearby as a stand-by rescue ship.

There was no sleeping that night. The troops

on the ground kept an eye out for an NVA attack that never materialized, in part because of the concentrated air support attacking the horseshoe ridgeline. We in the air stayed alert watching for potential enemy troop movements while also hoping against hope for a signal from our downed pilots and crew. The C-130 Specter gunships kept up a continuous stream of 7.62 tracers around the two bomb craters. Reports from the ground were steady, with a few NVA mortar rounds peppering the crater area. Kilburn relayed, much more calmly than before, "All is well, and keep up the air support."

Then, at 0500 hours, tactical air commenced striking the ridgeline, pounding away at the antiaircraft positions. Four sets of F-4C Phantoms dropped their bombs and fired their 20 mm guns at the ridgeline. There was something extra at work within these men; they had lost fellow pilots in the target area, so you can be damned sure they were delivering their ordnance with precision and attitude. The NVA antiaircraft fire ceased.

Forty-five minutes after the beginning of the heavy-duty tactical air assault, the four H-34s landed two by two in the bomb craters. The hunched Hatchet Forces personnel hurried to board the birds that would take them to safety. They piled in as quickly as humanly possible, knowing any delay could prove fatal. No one was left behind. Kilburn was the last to hop into a waiting bird.

A head count of the Hatchet Forces conducted

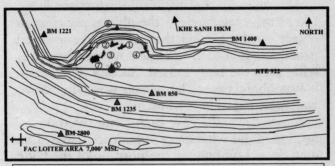

OSCAR-8, SECOND SECTION:
SKETCH OF THE CRASH SITES

① AND ② CH-46 TROOP CHOPPERS ③ F4-C PHANTOM ④ A-IE SKY-
⑤ HU-ID GUNSHIP ⑥ H-34 SOG CHOPPER RAIDER
★ ⑦ BOMB CRATERS WITH HF TROOPS DEFENDING

at Khe Sanh revealed the extent of our losses. Our people were littered throughout Oscar-8— fourteen U.S. pilots and crew members, ten indigenous SOG Hatchet Force members, and the three U.S. SOG men. The three missing SOG men were Sergeant First Class Charles Wilklow, Sergeant First Class Charles Dexter, and Laney, who was reportedly killed and his body not recovered. We continued our exhausting search. We flew at least twenty hours a day over the next two days, searching and searching, like men scanning the ground for gold, and found nothing.

On the morning of day four, June 8, 1967, Alexander and I canvassed the western side of the horseshoe ridge. I noticed an unusual color and lump on the ground directly beneath the right wing of the 0-2 aircraft.

I bolted upright and yelled, "There's a fucking body down below, hard right!"

Alexander pivoted the bird off the right wing. I kept my eyes fixed on the object, noticing an object on the ground that looked like a signal panel. We looked closer and saw the panel was lying atop the body's chest. Success. Goddamn, it was about time. The person was lying down and appeared to be a white male.

Alexander and I gave each other the thumbs-up. I then did what every experienced air search man does: I pinpointed the position of the body by some landmark. In this case, it was a group of tall trees in the vicinity that identified the particular spot. Alexander had done the same thing, pinpointing a different landmark for his target.

A mad rush of excitement surged through my body. Pure, unadulterated adrenaline cut through the exhaustion and frustration of the past four days. I called the Khe Sanh base and requested an H-34 rescue bird rigged for rope rescue be launched immediately. I requested two SOG men, but just one—Staff Sergeant Lester Pace—was available. Pace was a large black man with hellacious physical strength. He also tended to be a loner, and surly, which had created tension between the two of us in the past. But the past was forgotten amid the immediacy and importance of this mission. Nothing mattered but getting that body off the ground, and if Pace was the only one available for the duty, then Pace it was.

I briefed Pace over the FM radio, giving him the instructions needed to complete the rescue mission. "Pace," I said, "you're going to have to listen to me on this. Here's what you're going to do. You're going to rappel from the H-34 to the side of the man on the ground. Wear your extraction rig down, and carry an extra rig." This extra rig—called a Stabo rig—allowed for a man to be extracted from a hostile area, and then dangle by climbing rope under the helicopter until a safe LZ can be found. "When you reach the man, Pace, hook him and yourself to the extraction bar at the end of the climbing rope, under the helicopter, and give the pilot the thumbs-up for liftoff as soon as both of you are securely fastened."

There was a pause on the other end as I allowed these instructions to sink in. I wasn't asking Pace to complete a conventional rescue; this man on the ground needed to be extracted as quickly as possible, or else Pace and the H-34 stood a good chance of being shot from the skies.

When Pace didn't offer anything, I continued, "Pace, don't come up without the man. Beware of booby traps, and if you see any booby traps on the body, give the signal to take you out."

I was asking Lester Pace to complete something that was nearly impossible. His overwhelming silence made it clear he understood as much. He finally mumbled something about not completely understanding my orders. I repeated the instructions, and Pace finally "rogered" the message.

We continued to fly over the figure on the

ground, awaiting Pace and the H-34. Suddenly, as if jolted awake, the person sat up and began waving the rescue panel toward us.

"It's an American," I said to Alexander. "He just moved. He's standing up, and he has a panel. Give me some altitude now."

This person—I had no idea who it might be— was now holding the red signal panel with both hands and flashing it. Alexander gunned the 0-2's engine to let the man on the ground know we had him in our sights. We had no other means with which to communicate this fact.

Obviously, we couldn't loiter in this area too long. We gained altitude again as the H-34 came into view. I let Pace know the situation below, and I let him know both loudly and clearly, "He's alive, Pace! He's alive! Get your ass down that rope and save him. Can you see him?"

"Roger that," Pace replied, the anticipation apparent in his voice.

The H-34 hovered over the position, and Pace began his rappel. Hanging himself out there for all the NVA to see, Pace lowered himself to the ground, hooked up the downed soldier, and gave the thumbs-up to the pilot. Pace did a fantastic job, just perfect, and Alexander and I exchanged smiles as Pace and Sergeant First Class Charles Wilklow, temporarily MIA in Oscar-8, lifted out of the rescue area and flew west away from the horseshoe ridgeline.

Wilklow had been seriously wounded in action with a gunshot wound to his leg. The Special Forces medics told me later that Wilklow's wounds had become a feasting ground for jungle maggots. Those maggots were a lucky break for Wilklow; they ate away the dead flesh, helping to save his leg.

I searched Oscar-8 for five more days. I searched and searched and then searched some more for Sergeant First Class Billy Ray Laney, Sergeant First Class Charles Dexter, the fourteen MIA pilots and crew members, and the indigenous personnel. Again, no joy. No sightings, no radio transmissions. Nothing.

The rescue of Wilklow had given us hope, but it was quickly dashed as the days wore on. Wilklow's rescue was one of those miracles of war, hope emerging from a seemingly hopeless situation.

At the end of the Vietnam War, some of these MIA personnel were released from a prisoner-of-war camp in North Vietnam. SOG Sergeant First Class Charles Dexter was reported to have died while captured.

The saga of Oscar-8 ended unsatisfactorily, but at the time the potential gain was worth the risk. If we had killed or captured General Giap, the war would have ended or been shortened. Oscar-8 was a noble failure, and I tip my beret to the men who participated in the mission. I hold a special place for those we failed to rescue, including Billy Ray Laney, Charles Dexter, and the U.S. pilots.

In the late 1990s, a portion of Laney's body was found and returned to his daughter, Vickey

Laney Workman, in Minnesota. In 2001, I invited Vickey Laney Workman to the Special Operations Reunion in Las Vegas. She and her husband were in attendance as we stood to honor her father, who died when she was just six years old.

For thirty-three years, Vickey kept faith that her father would be returned to her from the jungle of Laos and Oscar-8. As Vickey stood before the men of SOG, I sat beside her and quietly said, "Say a few words to these men and their wives, please, Vickey."

The vast banquet hall was absolutely silent as the SOG men looked at this tiny lady who had kept faith for so long. Tears ran down Vickey's cheeks as she told the story of burying small bits of her father's bones at the veterans' memorial in Huntsville, Alabama. While the room remained silent, Vickey Laney Workman related, in her tiny voice, how she had placed a few of her father's bones in a full-dress green Army uniform, with paratrooper boots below, ribbons and sergeant major stripes above. On top of the uniform she placed the Green Beret her father was so proud to wear. This was the arrangement of the coffin bearing the remains of Sergeant Major Billy Ray Laney.

Through the tears, she said, "I thanked God for returning my father to me and my family, for we loved him so very much. We are very proud of my father. We are so happy to have him back at home with us."

Her quiet voice resonated through the vault-like silence of the vast room as she related this American story of loyalty and love. Six hundred SOG veterans sat in that room—six hundred hardened, combat-tested, and battle-scarred men who had seen some of the worst humanity had to offer as they fought the battle of the Ho Chi Minh Trail. There was not a single dry eye among them.

CHAPTER 4

On the morning of February 2, 1970, I was ordered by the commanding officer of MACSOG's Operation 35 to report immediately to his office in the Saigon area. I was given specific instructions: Report to the CO's office alone, with combat gear in hand. I lived for these words. These words told me a mission of utmost secrecy was imminent.

In the CO's office I was met by Lieutenant Colonel John Lindsey, who immediately outlined a situation he had witnessed from the air in a place called Ba Kev, thirty-two kilometers inside Cambodia, due west of the Duc Co Special Forces camp. From the back seat of a U.S. Air Force Cessna 0-1 Birddog, Lindsey had seen old people and children, their emaciated and malarial bodies lying along the fringes of a dirt airstrip. He saw a ragged collection of Cambodian soldiers and their families waving toward his plane, as if offering themselves up to it.

Standing forlorn on a patch of earth that had been scraped out of the jungle, they stared skyward and waved, imploring the plane to land on the base's rudimentary airstrip.

Lindsey told me the camp had been used by the NVA as a way station along the Ho Chi Minh Trail before the coup that removed Prince Norodom Sihanouk from power in Cambodia. The change of leadership—from Sihanouk to the U.S.-sponsored Lon Nol—kept the NVA from using the outpost, but it also ended resupply air drops to the people living in Ba Kev. These people were political orphans, and Lindsey saw an opportunity for us to take advantage of the situation.

He offered me a remarkable mission: Take a group of three interpreters and two other SOG men into Ba Kev and turn the outpost into a functioning camp that could be used to launch offensive strikes against the NVA across thirty-two kilometers of dense jungle to the Ho Chi Minh Trail. This top-secret mission would take American troops farther into enemy territory than ever before, and it would call for us to establish contact with a large group of armed people who were sick, hungry, and possibly unfriendly to our cause.

Would I do it? Lindsey asked. I saluted crisply with a big Texas grin creasing my face. Would I do it? Are you shitting me? I'd kill to lead such an operation.

An unconventional, spur-of-the-moment, cross-border operation deep inside enemy territory was exactly what I advocated since entering SOG. My entreaties were always greeted with blank stares and deaf ears, but now I was hearing wonderful news. I was ready to go that minute.

This is the first extensive chronicle of a remarkable mission. SOG files were declassified in 1995, but none of the few participants in the Ba Kev operation have shared their story until now.

The interpreters were chosen by headquarters, and I quickly rounded up the two SOG men: communications specialist Staff Sergeant Dennis Motley and demolitionist/engineer Sergeant First Class Charles Smith. Motley had been on two cross-border missions, and the uniqueness of this mission made experience imperative. Smith, a thirteen-year Army veteran, had seen it all and bore the look of a man who didn't expect to see anything new. He accepted the assignment with a shrug and a nod.

Like any great Special Ops mission, this one filled me with a combination of excitement and uncertainty. I have to admit my sphincter was a little tight as the H-34 chopper lowered into Ba Kev and I watched from my perch above as the people of the outpost walked toward us with wide-eyed wonder. We were looking down at a force of unknown size, a force known to possess a thousand-plus guns. We had no idea of their loyalty, but at the very best these folks were not NVA.

And the very worst? I didn't even want to think about that.

As the bird eased onto the ground, we didn't
know whether we would be greeted with open
arms or shot on sight. Both, it seemed, were distinct possibilities.

As we disembarked from the helicopters, we
were heartened by the lack of gunfire. I cau-
tioned the team in advance to smile and walk
slowly toward a predetermined spot—a shed on
the west end of the airstrip. This whole scene had
an otherworldly feel to it, as if we'd been
dropped onto a distant planet as some sort of ex-
periment.

I scanned the faces and uniforms until I spotted a man I took to be the ranking officer, a man
introduced to me as Colonel Um Savuth. This
man was in his midfifties, with a sharply lined,
weathered face. He walked slowly, with the aid
of a cane, and stood perhaps five feet, four inches
tall. He had a deep scar near his hairline that
looked suspiciously like it was caused by a gunshot wound. His skin was markedly yellowed,
which told me the colonel was most likely suffering from jaundice caused by malaria. As quickly
and innocuously as possible, I scanned his person
and saw he was armed with a Chinese pistol in a
leather holster attached to a French pistol belt.
His uniform, though worn, was clean. He held
himself as tall as possible, and my first impression was that I had come across a furiously
proud man.

In my best (not great) military French, I
greeted Colonel Savuth. As is customary when
dealing with Asian military commanders, I pre-

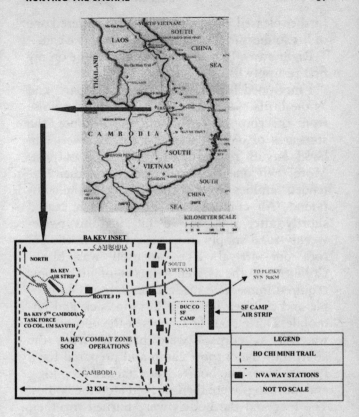

sented him with a greeting letter from Chief SOG. After a couple of minutes of awkward and idle chitchat, I presented the colonel with several small gifts meant to establish our friendliness. One of the gifts was a case of Budweiser, which really caused the old boy's eyes to light up. He disregarded its warmth, opened a can, and took a healthy swig. This was a good sign, and I was further pleased when Colonel Savuth ordered one of his soldiers to pass each of his officers a can.

Right away, I knew beer was sure to be on my first resupply list.

I motioned for the helicopters to depart, and as the birds rose over the treeline and the noise from the rotors receded into the distance, the starkness of our isolation hit me. Budweiser in hand, Savuth related the history of his outpost. More than one thousand troops and their extended families lived in the vicinity of the compound. All contact with Phnom Penh ceased shortly after the coup d'etat, and his twice-weekly food supply drop ceased four weeks before our arrival. The situation was bleak. Colonel Savuth, as well as many of his troops, required immediate medical attention. Old people and children lined the airstrip, sick and immobile. During our first hours at the outpost, we watched as corpses were brought from the airstrip to the outpost camp and prepared for a funeral pyre. The bodies were stacked atop logs, and more logs were then stacked atop the bodies. The logs and the bodies were burned together. When nothing was left but ashes, they were spread in the jungle.

The NVA had used the outpost as a way station. Soldiers who were wounded or ill would stop at this depleted medical aid station for a few days to gather their strength. While there, the troops would take any food or medicine they felt they needed, regardless of the needs of Colonel Savuth's unit. This proud man was visibly disturbed by this practice. Clearly, he was encouraged by the prospect of us replenishing his depleted or nonexistent supplies of food and medicine. This was a man who wanted to take care of his people and seemed willing to work with us to achieve that goal.

After we dispensed with the formalities, Colonel Savuth looked at me with his yellowed, tired eyes and asked a simple question: "What can you do to protect us?"

I was ready for this. We knew our presence would spark the attention of the NVA, and the colonel did not wish the wrath of the NVA to descend on his unit. I had alerted the C-130 airborne CP, call sign Hillsboro, to our presence in the Ba Kev region. We were prepared to answer the colonel's question in an emphatic, spectacular way.

I turned to Colonel Savuth and pointed to the surrounding mountains. "Pick a spot," I told him. "Somewhere at least four hundred meters away, where you would like to see a display of U.S. might."

Savuth pointed to the east, approximately eight hundred meters from the airstrip, to a wooded knoll. "Do you see that?" he asked me.

I related an approximate eight-digit grid coordinate to the USAF pilot flying with Lieutenant Colonel Lindsey and requested a "Sky Spot"—two 250-kilo high-explosive bombs to be dropped on the coordinates of the knoll immediately.

By now many of the Cambodian soldiers were

lining the airstrip, sensing the nutty Americans were up to something unusual. Within five minutes, two fighter aircraft far above, neither seen nor heard by the Cambodians, dropped their ordnance directly on the target. The bombs were not visible to any of us during their descent, so the explosion was a riotous surprise that took the breath away from the Cambodians, including Colonel Savuth.

As the mountain erupted in flame and smoke, a huge roar rose from the soldiers. They stood along the airstrip and clapped, laughing uproariously and reliving the moment with those around them. I felt a wave of giddy relief flood over me. I stole a satisfied glance at Colonel Savuth. The colonel blinked his eyes, smiled, and threw down some more Budweiser. I thought, *Damn, this mission's got a chance.*

The NVA would attack. There was no way in hell the NVA would permit a U.S. presence in this area without attempting to take us out. The Bodes' defenses were not defenses but merely living areas. Our SOG team was sent here to help this Cambodian outpost survive and then initiate offensive combat versus the NVA. If we were successful, we could launch small-team patrols at the Ho Chi Minh Trail from the east, a strategy that would take pressure off the overworked and overbombed Duc Co Special Forces camp, just across the border of South Vietnam in the II Corps tactical zone. So if we were going to be living here, with the NVA breathing down our necks, we had a hell of a lot of work to do. This new home was definitely a fixer-upper. We had to build defenses, establish communications, build morale, and treat the medical needs of the troops and their families.

Our first order of business was Colonel Savuth's health. As I spoke with him on the afternoon of our third day in camp, he became too weak to continue and had to return to his headquarters to rest. This old boy needed medical care soonest. For morale purposes, I needed to make his evacuation a top priority. We could not let Colonel Savuth die in the camp. His death would be the end of the operation, and if we didn't get him some help within twenty-four hours, he was likely a dead man.

The sweet sound of SOG helicopters dopplered into the outpost on the morning of February 5, the third day of our occupation. They carried a three-man Special Forces medical team, with a load of medical supplies. As soon as the choppers landed, the lines began forming. Soldiers, old people, children, pregnant women—they all lined up around the camp, awaiting lifesaving medical treatment from two SOG physicians and a Special Forces medic. These men were treated like messiahs; the people were so weak and sick they could barely stand in line, but they waited patiently while the medical team spent eight hours treating about five hundred people for malaria and other ailments.

One of the doctors evaluated Colonel Savuth,

who by now was almost totally confined to his hut and the wooden rack that served as a bed. The colonel ran a high fever, and he was shaking uncontrollably from the chills. After the doctor finished his evaluation, he approached me with a grave manner.

"He needs to get out of here," he told me. "He needs to get out of here *now*. Any delay could be fatal."

Colonel Savuth understood his medical situation but was reluctant to depart. "I need to stay," he told me. "I can't leave now."

"Colonel," I told him, "the truth is, you won't last two weeks if you aren't treated." After some deliberation, he agreed to go. As the day's last light faded over the mountains, Colonel Savuth officially turned over the outpost command to Major Um Thant and was then placed aboard one of the SOG helicopters by the exhausted members of the medical team. As the H-34s departed to the east, each of Colonel Savuth's regimental officers stood at attention on the airstrip, saluting their proud old colonel as he went off to be treated secretly by Special Forces doctors at the trusty 8th Field Hospital in Nha Trang, South Vietnam.

With the medical needs of the outpost beginning to be met, we organized air drops every other night; the extent of these drops left the Cambodians awestruck. Major Thant wore a permanent grin and continually shook his head in amazement as I detailed for him the contents of the shipment—a three-day supply of rice and rations for three to four thousand, rolls and rolls of razor wire for defense, one hundred pairs of engineer gloves, six chain saws, Claymore mines, fifty-five-gallon drums of gasoline and diesel. These people had never seen anything like it.

We watched as the Stockholm syndrome took effect. If you treat an adversary well, he eventually becomes your friend, and we worked this principle to the fullest in Ba Kev. As I looked around at the faces of the Cambodian soldiers as they marveled at the air drop, I knew they were beginning to look upon us as their saviors. This was exactly the position we wished to occupy in their minds.

Around-the-clock work crews made the outpost able to defend itself to a minimal degree, and perhaps withstand an enemy attack. The humanitarian aspect of our mission was proving to be an unqualified success. Now it was time to implement offensive actions, the main reason for our unlikely presence in this remote Cambodian outpost. In other words, we had held up our end of the bargain. These people were healthier and safer than they were before our arrival. Now it was time for the Cambodians to return the favor.

On February 12, our ninth day in Ba Kev, we initiated local small-unit patrols. These patrols consisted of three- to four-man teams that were launched repeatedly, on foot, from the outpost position toward the Ho Chi Minh Trail. The purpose of the local patrols was not to destroy any Charlies but to give the commander an early

warning regarding approaching enemy. The
teams were taught to sit in ambush at natural av-
enues of approach, then immediately report any
enemy movement or activity.

Armed with an AK-47 assault rifle and five
magazines of ammunition, each man wore green
jungle fatigues with the pant legs attached to the
ankles with black tape. This allowed for quiet
movement through the jungle. Each also wore a
pistol belt with two canteens of water and carried a signal mirror and panel and a small compass. Selected members carried an M-79 grenade
launcher.

Two days after the first local patrols were
launched, a four-man team in an ambush position noticed a North Vietnamese soldier, walking alone and heading directly for them. This
NVA soldier walked with no possible reason to
believe a four-man patrol composed of Cambodian soldiers would ambush him several kilometers east of the Ho Chi Minh Trail as he walked
with his AK-47 slung over his right shoulder. But
walk he did, dead straight into the teeth of the
ambush. He was hustled out of the brush and
back to the outpost. And with that, the siege of
Ba Kev claimed its first enemy capture.

Our proud and happy patrol team rolled into
camp with a young, sick, and scared NVA named
Nguyen Van Dong, a nineteen-year-old soldier
who had volunteered for military service just
three months earlier. When I asked him his reason for being so near our position, he said, "My
leader sent me here to seek medical attention."

Apparently Van Dong's unit had not been informed of our presence in Ba Kev, or of the abandonment of this outpost by the NVA. Unsure of dates and slightly delirious from his sickness, Dong said he believed he had been marching with his unit for about four weeks before being sent off to seek treatment. He was just a young kid, a grunt who was scared shitless and sick out of his mind with malaria. He was no use to us, so we arranged for him to be transported to Duc Co for further interrogation and treatment.

This prisoner snatch didn't provide us with any valuable information, but it was a big development in Ba Kev. I told Sergeant Major Uk Saddan, "The men involved in this snatch should be rewarded." So that afternoon, after Dong lifted off for Duc Co, we formed a portion of the regiment and blasted the patrol members with the glory gun. Saddan presented each of the men with the equivalent of $20 U.S. in Cambodian riels. As these soldiers were hailed by their comrades for a job well done, they were as proud as four men can be.

I allowed myself a brief moment of pride before snapping myself back to reality. Amid the revelry, I reminded myself that Charlie wasn't going to allow our presence here, in direct interference of his march south, without reciprocating.

We didn't have to wait long.

I had learned, out of necessity, to sleep with one eye and ear open. I didn't sleep much, and I didn't sleep soundly, which meant I didn't miss

much when it came to nighttime noises. At 2
A.M. on February 15, one day after the capture
of the NVA soldier, I awoke to the dreaded thuds
of NVA mortar rounds being dispatched from
their tubes. They were on their way, and within
seconds they paid us an unwanted and very loud
visit.

As I shook myself awake and hurried to alert
the rest of the outpost, I took a split second to
realize Charlie had finally become concerned
with our presence.

I sounded the alarm at the top of my lungs:
"Incoming mortars!"

We were all "scramblers" at this point, doing
our best to alert the entire camp as quickly as
possible. Some of the Cambodians were screaming "Faire attention!" and everyone was scurrying like mad. Fear has a way of igniting the speed
in even the slowest of humans, and within seconds everyone in the Ba Kev outpost—including
the families who resided on either side of Route
19—was buttoned up and ready to defend our
position.

I dived into the trench in our team's bunker,
along with Smith and Sergeant First Class
Melvin Hill, who had arrived the previous day to
take over our communications net from the overworked and overwhelmed Motley. When he left
Saigon, I told Hill I had a good deal for him, and
now he was muttering under his breath and cursing me as NVA mortars landed all around us.
"Goddamnit, Sergeant Major, you son of a
bitch," he said, "if it weren't for you and your

'good deals' I'd be in Saigon right this minute, chasing some young thing. Instead, I'm out here in this fucking hole in the ground."

As the mortars began to slam against the southwest side of the main camp, I told Melvin, "Shut the fuck up and get on the radio. Get us a fucking Specter gunship out here."

Hill stopped his whining and said, "I already have. The gunships are on their way." The immediate and affirmative response from Hillsboro sounded like God's voice from the heavens. At the same time, one of our local patrols, three kilometers to the east of our outpost, reported sighting two separate NVA mortar crews. Our patrol estimated the enemy mortar crews were less than three hundred meters to the north of its position. We did not order the patrol to fire on the enemy, for good reason: The patrols were too small, and their position too precarious, to engage this unknown force in combat.

Accompanied by one of our interpreters, I sprinted to the 82 mm mortar position and leaped into the mortar pit with the Cambodian crew. The interpreter told the patrol nearest the enemy to keep their asses down while we fired an HE "marking round." We needed the patrol to keep an eye on this round, in order to adjust our range and aim as needed. I crossed my fingers in the hope our marking round didn't land on the collective heads of our four-man patrol. It didn't, but it didn't reach the enemy, either. According to the feedback from the patrol, our round landed about five hundred meters short of the en-

emy position but directly on line. I had the tube
adjusted and then fired several rounds. These
rounds landed and detonated in the area the patrol had heard and seen the enemy fire. The patrol reported back to us that they had heard a
good deal of branch-breaking and chatter from
the general vicinity.

"The bad guys just might be scattering," I
said. I also hoped they had some shrapnel in their
asses as they scurried through the darkness of the
jungle.

Our troops then fired from a second mortar
position, aimed at the spot where the NVA B-40
rocket originated. (They took an immediate liking to the firepower, which builds morale but
wastes ammunition.) This mad, frantic sequence—maybe not more than a few minutes—
gave the regimental troops a good idea of what
combat is all about. And despite their understand-
able lack of discipline—these lads sure enjoyed
their weapons—I was proud of these men for an-
swering Charlie's call with a few lethal messages
of their own. They didn't seem bothered at all by
the twist of fate—two months before they were
housing and feeding these same soldiers they
were now attempting to blast out of the jungle.

At first light the next morning, the Cambodian
troops along the berm in each of the camps chat-
tered like a flock of magpies. The sound was
continuous and enthusiastic as they relived every
second of the battle. Melvin Hill and I smiled as
we watched these men experience their first combat high.

Our casualties were limited to two killed Cambodians and three wounded by a B-40 rocket in Support Camp Alpha. The Cambodians cremated the dead immediately by placing the two bodies in a ceremonial pyre like the one we first witnessed upon our arrival at the outpost. Again, we watched with blank faces and worked to remain stoic. This was Hill's first exposure to this unusual practice, and the look on his face was a mixture of curiosity and revulsion. "If that's the way they do it," I said, "let them have at it." Melvin muttered in half-hearted agreement.

During our after-action discussion, I complimented the outpost commander on his lads' fast and proper response to the enemy actions. As I let the men know how proud I was of their first foray into combat, I thought to myself, "These guys just might turn out to be pretty damned good combat troops." I almost laughed when I thought it, because two weeks before I never would have guessed those words would cross my mind.

The NVA didn't want to have to deal with us out there in the boondocks, but we gave them no choice. We were picking up the pace of our patrols and had shown the ability to withstand the NVA mortar attack. We were a nuisance, a pesky pain in the ass Charlie couldn't afford to ignore.

On February 16, Lindsey contacted me from the air and requested my presence—along with the French-speaking SOG captain James Spo-

erry—in Pleiku to meet with Lieutenant General
Lu Lon, the commander of South Vietnam's military Region II. I was asked to bring along one of
the outpost's regimental officers, and together
Cambodian Lieutenant Um Ari and I boarded a
SOG-assigned H-34 headed for II Corps headquarters in Pleiku.

General Lon informed us of reliable information indicating the NVA's intent to overrun the
outpost within the next few weeks. He asked a
simple question: "Would the Cambodian regiment prefer to remain in the Ba Kev outpost, or
would the outpost battalion rather move to
South Vietnam?"

This was a watershed moment: We had reached
the point where our offensive actions had forced
the military honchos, both U.S. and South Vietnamese, to devise an exit strategy. Directing his
words to Ari, General Lon said, "The South Vietnamese government has authorized me to offer
sanctuary and refuge to the entire regiment, including dependents. This can happen if you agree
to convoy into South Vietnam and turn over your
arms and equipment to the South Vietnamese
government." An official offer was placed in
writing and addressed to Colonel Um Savuth,
who was completing his medical treatment.

Lindsey was his usual stoic self during the
meeting, but when it was finished he pulled me
aside and gave me direct orders from Chief SOG.
We were to continue to prepare defenses. We
were to continue to step up offensive operations,

taking prisoners when possible. We were to be prepared to train and exploit the Cambodians in offensive helicopter raid and long-range patrol operations.

Exit strategy? As far as SOG was concerned, exit strategies were someone else's concern.

On February 19, our sixteenth day in Ba Kev, we welcomed back our old friend Colonel Um Savuth. He appeared to be fit and in good spirits, completely recovered. He held a meeting with his officers and then summoned me and SOG captain James Spoerry, a French speaker who had recently joined our team.

Savuth patted me on the back and said, "Billy, your team saved my life."

The colonel then presented me with two beautiful, deep blue Chinese Tokarev pistols. I was deeply humbled and unsure how to accept such a wonderful gift, but I did accept them modestly and gratefully.

Colonel Savuth waved off my gestures of gratitude. Addressing the prospect of departing Ba Kev, he said, "Billy, we will evacuate Ba Kev when you and Captain Spoerry tell me it is necessary."

Colonel Savuth went right back to work, and his troops continued to show their respect for him and his methods. I frequently watched as a soldier was brought before him for some infraction of regimental rules. This gave me a good look at why his soldiers were so respectful and disciplined in his presence.

The colonel was partially crippled and walked with a cane. He would sit outside his shack in a straight-back chair as the squad leader recited the soldier's transgression. The colonel then looked dead in the eyes of the soldier, who attempted to hide his fear by standing as straight as humanly possible. After a few moments of uneasy standoff, Colonel Savuth would raise his inch-thick cane and whip it across the shoulder of the offending soldier. The number of cane whips was determined by the infraction. For an old, sickly man, the colonel could wield that cane. Believe me, these weren't love taps.

I never saw a soldier whimper or cry out in pain. These men never, ever made a sound. On numerous occasions the soldier's eyes would glisten with tears as he stood upright and girded himself against the force of the colonel's blows, but silence ruled. There was no sound. When the colonel was finished, he would simply dismiss the soldier and send him on his way.

I was fascinated by the colonel and his methods. Late one afternoon, shortly after his return from the medical leave, Colonel Savuth asked me to sit with him outside the shack while he drank a few cans of Budweiser. From the moment I met him, I had been curious about the wound to his forehead, and I figured I had reached the point where I could finally ask him what happened.

"Colonel, mind if I ask you how you received your wounds?"

Colonel Savuth took a big swig of Budweiser and told me an incredible story.

Some years back, the colonel and a major friend were drinking French "bier larue" in an establishment in the capital city of Phnom Penh. This was a potent beverage, and drunkenness—as it tends to do—led to argument.

The point of contention was marksmanship. More to the point, who was more accurate with a pistol. Both of these men considered themselves expert marksmen, so they decided to summon the ghost of William Tell and settle the issue.

The major paced off about ten yards, turned and faced the colonel, and balanced a beer bottle on his own head. The colonel promptly and confidently aimed his pistol and shot the bottle clean off the major's head.

The process was repeated with a beer bottle resting atop the colonel's head. The major missed the bottle and shot Colonel Savuth in his forehead, near the hairline. The old boy did not die, but he became crippled when the bullet creased his brain. He did win the contest, though, which wasn't much consolation.

I sat dumbfounded as I listened to this story, attempting to hide my shock at the ridiculous events that had caused the colonel to become partially paralyzed. I imagine the shock was readily apparent on my face, but the colonel didn't seem to notice.

"Did the major's poor marksmanship make you mad?" I asked.

The colonel laughed at this question and dismissed it with a wave of his hand. "Oh, no," he said, again swigging from his beer. "We're still good friends."

It was a memorable stay in Ba Kev, the most fulfilling operation of my SOG career, and with it came many unforgettable moments. On March 1, nearly a month into our occupation, I was in a hurry to move an injured American soldier—a SOG man who arrived well into the operation. This was during one of the many periods in which increased chatter led us to believe the NVA was preparing a final assault on our outpost. Obviously, concern over such an attack dictated much of our activity and influenced our thinking.

I put out a call via Emergency UHF to wheedle and cajole someone into picking up the wounded soldier. An unidentified U.S. aircraft responded immediately, and I asked the pilot what type of mission his bird was conducting.

"On a goddamned milk run," he said, meaning he was ferrying troops or supplies around South Vietnam.

"Does your radio compass give you a direction for my signal?" I asked, testing him to see if he was aware of my unusual, top-secret location.

"Roger that," he said. "You are to our west on a 275-degree azimuth. What can we do for you?"

I emphasized my location by saying I was approximately thirty-two kilometers west of the

Special Forces camp at Duc Co. The pilot apparently did not grasp that "thirty-two kilometers west of Duc Co" translated into "thirty-two kilometers inside Cambodia." This was fine with me. Happy to be diverted from his milk run, the pilot said, "Roger that, we are on the way." He told me he had blanket permission to land for the med-evac of any U.S. personnel.

Soon I heard the drone of a C-123 Provider. I looked up and saw the camouflaged two-engine workhorse appear to the east. The C-123 looked a little like a bodybuilder—a thick trunk attached to a couple of stumpy, wide wings. It was a mule of the air, created to carry cargo and troops reliably and unspectacularly.

The pilot made a low pass over the Ba Kev airstrip, and I wondered what was going through his mind as he guided that big bird over our weed-infested dirt patch.

"What's the condition of that strip?" he asked me.

I was afraid he would ask that. "Oh, it's just great," I said, hoping rather than knowing. "The center's fine for landing, but there's some soft earth outside thirty feet of the center. Keep her in the middle and you're fine."

"Roger," he replied.

He eased that bird in for landing. After a short roll, I could tell that big-bellied, barrel-chested C-123 was laboring to roll much farther. Its landing gear burrowed into the soft earth like a frightened gopher, and the plane ground to a halt.

I checked my watch and noted the time—
1030—figuring I might need it for any reports to
come. My pilot friend hopped out of the plane,
visibly shaken. His crew followed. They took a
long and critical look at the underbelly of the
bird. Even I—ground man that I am—could tell
the aircraft was stuck in the soft earth.

The pilot looked at me, eyes huge. He was ag-
itated, trying to hide the commotion inside his
head. Quietly, with something approaching
dread, he asked me, "Where in the fuck are we?"

I smiled a little and stared at him. "Welcome
to Ba Kev International Airport," I said.

"Ba Kev? That's in Cambodia, isn't it? What
in the fuck are you people doing here?"

"Some ground combat versus the NVA," I said
with a shrug. "Same as everybody else."

The pilot paced around a little bit. Ran a hand
through his hair.

"Damn," he mumbled. "I'm sure in trouble
now. I'm stuck, and I'm in fucking Cambodia."

I felt for this young Air Force pilot. After all,
he and his crew had answered my call and made
a brave effort to get their bird on the ground.
Now my goal was to expedite their departure
from our territory, and pronto.

"What will it take to get you unstuck?" I
asked.

"A lot of digging and some JATO bottles."

When attached to the underbelly of the plane,
Jet Assist Take Off provides a rocket boost that
increases thrust and lift. This, combined with the
C-123's propeller power, should get him off the

ground, over the treeline, and on his way.

One great advantage of this mission was SOG's complete noninterference in the proceedings. We operated on a no-questions-asked basis, mainly because the mission was paying off for SOG with MACV and the Nixon administration. At the time, it might have been one of the few successes of the war. So when we called for JATO bottles to be delivered from the Special Forces C team in Pleiku, by 4 P.M. a SOG H-34 helicopter was landing at Ba Kev with the bottles on board.

Our troops had been digging like hell, trying to carve a path for the C-123. We were fighting daylight, knowing an overnight stay of this bird on our property was certain to attract the NVA, who would undoubtedly destroy it. After we completed a two-hundred-meter path—as long and wide as the bird's wheels—the crew chief attached the JATO bottles to the aft portion of the aircraft while the rest of us stood by and hoped.

Once in place, the pilot gave us a thumbs-up and revved the two engines, with the brakes on. I said a little prayer to The Man, asking him to get this baby off the ground and over the trees and into the wild blue yonder.

For my sake.

Please.

The JATO bottles fired at the same instant the pilot released the brake system. The bird practically leaped upward to such a degree that the slanted tail almost struck the runway. The aircraft reached a height of about two hundred feet, dipped its nose slightly, and gained enough air speed to climb up and away from Ba Kev.

The pilot, relieved, got on the radio and said he was delighted to have landed in our unusual location, even though the injured American was med-evaced on the H-34 that brought the JATO. As he banked left and headed east toward South Vietnam, the pilot bid our team adieu. His last words were loud and clear. "Sergeant Major," he said, "please don't call me again."

As expected, our offensive actions quickened our departure from Ba Kev. We understood the situation—the more successful we were at harassing and disrupting the NVA, the sooner we were going to be forced to leave. The enemy would deal with us—and deal with us severely—when he felt enough was enough.

During the final two weeks of March, we began receiving harassing fire from the NVA more frequently. Two more NVA stumbled into the Ba Kev zone, sick as hell with malaria. During ambushes, we captured a couple more NVA and sent them back to Duc Co. Our patrols killed twenty NVA soldiers. We were becoming more of a pain in the ass to the enemy, and our Cambodians received information that indicated the end was near. The word was out: The NVA was set to march against Ba Kev and eliminate it from the earth, period.

This message immediately reached the top of the chain of command—meaning the White

House, for sure. They decided enough was
enough. They ordered SOG to shut down the operation.

The proposed evacuation of Ba Kev's people
and material was breathtaking in scope. We were
to move the entire battalion and everything asso-
ciated with it—soldiers, wives, children, dogs,
cows, donkeys, goats, trucks, gear, weapons,
and whatever else was lying around. It would be
moved east, along Route 19, across the Ho Chi
Minh Trail and to the Duc Co Special Forces
camp. Once there, the South Vietnamese high
command would treat these people with respect,
confiscate the battalion's weapons, and relocate
everybody to the Cambodian capital of Phnom
Penh.

This was billed by the leaders of SOG as a humanitarian action, and a contingent of media arrived in Ba Kev to film the move of our fine
Cambodian unit. The media was not informed of
the circumstances leading up to the evacuation,
and Colonel Savuth was told to brief his troops
to remain silent regarding the presence or past
presence of an American unit in their midst. I
asked for an air cover plan from Operation 35
and received assurances that tactical air would
be within five minutes of Route 19 during the
entire movement eastward. The top-secret nature
of SOG also made it imperative for us to stay
away from cameras and reporters. When the media assembled, we disappeared and Colonel
Savuth took control.

We were ready to pick up an entire community and relocate it for its own good. We had to
wait and see whether the NVA would play along.

The trek began at first light on April 2, 1970.
Colonel Savuth, knowing the convoy must reach
the South Vietnamese border before nightfall,
herded it like a trail boss on a Texas cattle drive.
Two SOG H-34s arrived at dawn to transport
the U.S. team back and forth across the convoy.
The mass of humanity started out to the east,
along Route 19. Since the mechanics couldn't
coax many of the Russian and British trucks into
performing, many of the Cambodian family
members were forced to walk. The working vehicles were jammed with as many people as
physically possible. H-34s ferried the lame and
the old.

The NVA's spies were always at work. With
the arrival of media and the accompanying commotion, there was no way the enemy could have
missed the preparations for this move. We could
only conclude that the NVA high command had
spread the order not to harass the convoy.

Melvin Hill and I watched from overhead in
an H-34 that flew back and forth to provide protection to the convoy. As I looked down on this
ragged-assed unit attempting to wind its way
east and out of Cambodia, I was most proud of
Colonel Savuth. I watched him closely as he was
driven up and down the flanks of the convoy as it
moved east along Highway 19.

From my vantage point, this was the sight of a

lifetime, a revision of the biblical Exodus story. Moving excruciatingly slowly, a long line of mankind and all his earthly possessions, spread out over more than five kilometers, trudged its way to a new life. The goats and cows fell behind, but they continued their slow and steady march east under the direction of several young Cambodian children—the children of our soldiers.

Eleven hours after departure, at roughly 1700 that afternoon, the convoy reached the border with South Vietnam. Colonel Savuth ordered the trucks to be unloaded and sent back along the route to collect stragglers. Hot rice and chicken was fed to these fine people, and the older ones showed their appreciation by touching us, placing their hands together, and smiling their beautiful toothless smiles.

The next morning, at about 1000 hours, Colonel Savuth and I went inside the Duc Co Special Forces camp together. We spoke about the events of the past two months as he walked slowly, listing to his left side. The colonel shuffled proudly toward a group of waiting VIPs, then gave his best attempt at a crisp salute when he came before General Lu Lon. As I watched this poignant moment I wasn't prepared for what came next: I felt myself choking up, and before I knew it a tear was running down my cheek. Damn, I was proud of this fine man.

After the colonel finished with the VIPs, it came time for him to bid farewell to me and the team. After sixty-two days, this was bound to be a tough goodbye. My first order of business was to present the colonel with twenty cases of Budweiser. He accepted this gift without hesitation.

I shook his hand and looked into his eyes.

"*Dieu vous garde, mon colonel,*" I said.

He blinked hard and bid me adieu. He then placed his arm around me and, in English, said, "Good-bye, Billy. Thank you and your team."

We each took a step back and exchanged hand salutes. As I watched Colonel Savuth depart, hobbling along with his ever-present stick, I knew I would never forget this mission or this man.

The last known record of Colonel Savuth is from 1973. He is listed in the CIA's Order of Battle as being a one-star general at the time, which means the good colonel enjoyed a fine military career following our time together in Ba Kev. I never learned Savuth's ultimate fate, but I am confident of one fact: Right this minute, wherever he might be, this man is with his people, and he is leading them.

CHAPTER 5

I returned to the United States from Vietnam in December 1971, after a total of seven and a half years in the country. The war was my life—seven and a half years, eight Purple Hearts, one Silver Star. Those are the quantifiable accessories to war, mostly meaningless and unimportant. But how much hard work? How much sweat? How much blood? How many lives saved? How many friends lost in combat? Those have permanence. They provide the legacy of those seven and a half years.

President Nixon's policy of "Vietnamization" was turning over responsibility for administering and fighting the war to the South Vietnamese. It didn't require much in the way of gray matter to see that the United States was not only cutting all ties with the South Vietnamese but handing South Vietnam over to the North.

This was a bitter blow to me. So much hard work. So much blood shed. So many hopes

dashed. I had a difficult time bending my mind
around the idea that we were ending this war in
such a capitulating fashion. President Nixon's
Vietnamization speech came in 1969, roughly a
year before we began our epic operation in Ba
Kev. This awkward and newly minted word—
Vietnamization—immediately became the over-
riding doctrine and theme for all U.S. operations
in Vietnam.

In a nod to my longevity, I was asked to escort
the withdrawal of the 5th Special Forces from
Vietnam in 1969, shortly after I came out of War
Zone D. I was among the first Special Forces
men to enter Vietnam in 1962 when 5th SF arrived in the country, and by God when the calen-
dar flipped to 1972, I was still there. I suffered
some serious body damage and spent some serious time being put back together, but I kept
coming back.

I agreed to carry the colors, along with many
other of the war's old-timers. Those colors were
reactivated at Fort Bragg, North Carolina, and I
promptly returned to Vietnam, and SOG, as sergeant major of Recon Company for Command
and Control South. This was my position during
the Ba Kev operation.

After Ba Kev, I was reassigned to Recon Company Command and Control North. I led a very
large recon company, with approximately
twenty-three U.S.-led recon teams, with nearly
three hundred combat-tested Bru tribesmen
working alongside.

President Nixon and his administration lost

heart for the war, but SOG didn't. Perhaps alone among the military, we stepped up our operations. In July 1970 I was chosen to assemble volunteer recon personnel for High Altitude Low Opening (HALO) teams that would insert into the Ho Chi Minh Trail by free fall.

We adopted and embraced the idea of HALO inserts because our helicopters were being blasted out of the skies more and more often. The NVA covered every landing zone in the area. The enemy waited for our SOG CCN recon teams to arrive on helicopter landing zones west of the Ashau Valley and killed five complete recon teams before the teams could move quickly from the landing zone. This is a total of about forty-five men, fifteen of them U.S. Special Forces. You couldn't hide a helicopter—it was both visibly and audibly obvious. The NVA ability to keep SOG recon teams from the field of battle gradually solidified its control of the Ho Chi Minh Trail.

With no weapons to go against their movement inside the jungle, save for the intercept by SOG operations, the NVA was able to move hundreds of thousands of men and matériel to the south. That trail was the reason the North Vietnamese prevailed, and we tried our damnedest to shut it down. The job was simply too big.

We had to put our teams on the ground, and HALO was the only answer. It was drastic and unconventional—maybe even desperate—but it was our only hope. HALO was a perfect example of necessity being the mother of invention.

We instituted the HALO insertions—done at night, often in bad weather, from altitudes in excess of twenty thousand feet—to regain the stealth we needed to hit the ground undetected. These HALO teams inserted from November 1970 to August 1971 with mixed results.

It was harrowing, dangerous work, and it evolved into an invaluable tactic for today's high-tech warfare. We were the guinea pigs, the guys who had the balls and the freedom to use our ingenuity and fearlessness to attempt to complete the job. Now, more than thirty years later, there are so many examples of our creativity meeting up with advanced technology to produce spectacular results.

In Afghanistan, small-team HALO inserts were the backbone of the effort that disabled the Taliban in a matter of weeks. In Iraq, small-team HALO inserts had the same effect. We in SOG were pioneers, but our vision was purely accidental; our main objective was to get to the enemy in the most efficient way possible, with the best chance of survival.

The distasteful retreat from Vietnam was a pivotal point in my career. My orders were up in November 1971, and I discussed them with the Operation 35 commander, Colonel Roger Pezzelle. The orders were to return to Fort Bragg, so I needed to make a decision—once again—on whether to extend my tour of duty.

The colonel quietly told me, "Billy, the United

States has lost heart for Vietnam. We will be closing CCN and CCC, so you may as well go back to Fort Bragg while you have the orders. SOG is finished as of March 1972."

This fine former OSS officer was explaining to me that we were going to give the country to the NVA. As he did, I thought, "What a perfectly fucking shame. Goddamn, why did we allow our lads to die? What a goddamn travesty this is."

I did not articulate these thoughts to the colonel. We both felt the same way, and we were powerless to fight the decisions made in Washington. So I simply looked at Pezzelle and said, "I know it, Colonel. It's over."

As I stood there, I thought, *Damn. Oh damn. We begged so many times to take this battle to the North, to Hanoi, directly to Ho Chi Minh and Vo Nguyen Giap. We could have killed those two and finished this war years and years before.*

Again, I kept these thoughts to myself. I knew the best time for proper action was 1965. Rather than piecemeal replacements and half-assed actions, we should have fucking well taken the battle to the heart of the enemy. We could have finished that war seven years before the last battle was fought by our SOG lads.

Our administration did not have the stomach for such a vigorous plan of action, so what did we do? We fed soldiers to their death for ten years, just to walk the hell away. After fifty thousand American men and women KIA we retreated from the battleground, with nothing to show for so much bloodshed. I knew Colonel Pezzelle, a fine officer and an old soldier, felt the same way. We didn't need to discuss it.

Instead, I shook my head sadly and slowly. We looked each other in the eye. I saw tears in his eyes, and I know he saw them in mine. The war in Vietnam, for me, ended at precisely that moment.

When I returned to Fort Bragg and the 5th Special Forces Group (A) in December of 1971, I had to find out what an armchair sergeant major was required to do. It was new to me; I had never performed in peacetime as a master sergeant or sergeant major, and the thought of keeping rosters, chewing ass, and struggling to keep busy did not appeal to me. Not in the least.

As I was signing in to headquarters, the command sergeant major sent for me. When I reported to the man we called "SMAJ," he said, "Billy, you've been overseas for a heck of a long time. We have an excess of your rank (E-9), so you're being reassigned to Fort Devens, Massachusetts, to the 10th Special Forces Group."

"No way," I said. I respected the man, and his position, but no fucking way.

"I will not allow this to happen," I told him. "I'm not going to Massachusetts, and I'm not going to the 10th Special Forces. Peri-fucking-od."

I believe the command sergeant major received my message, loud and clear, but if not, I reinforced it a bit.

"I am going to the Pentagon to settle this," I said. "And if I can't settle it, I will retire from this army."

The command sergeant major gave me the OK, and I traveled to the Pentagon in January of 1972, where I was received by four-star General William Westmoreland of the Joint Chiefs of Staff. He recalled pinning on my sixth Purple Heart in 1965 in the field hospital in Nha Trang.

From the beginning of our discussion, it was obvious I was going to have to make good on my promise to retire.

"Billy, Special Forces has angered so many conventional generals that it will be lucky if it survives."

I didn't even argue with him. The message was clear. I shook hands with Westmoreland and said, "Damn, General, I'm going to retire. And what a fucking shame, for I've enjoyed every minute of the military."

I struggled to keep my emotions in check, a difficult chore that became even more difficult when I saw Westmoreland's eyes begin to tear up.

I initiated my papers for retirement when I returned to Fort Bragg. I knew I was going to be lost with no war to fight, no men to depend on me for their lives, no pilots to rescue from the clutches of the enemy, and no fine SF men around me.

When my papers were approved, the command sergeant major of Special Forces asked me what I wanted to do for my retirement.

I thought for about five seconds.

"SMAJ," I said, "I want to free fall into the parade ground. I want you to jump with me and several of my friends. Then afterward I want you to give the men the rest of the day off, so we can drink some beer."

The command sergeant major ran my unconventional request up the flagpole and permission was given. At age forty-three, I jumped into my retirement.

When I landed, all of Special Forces assembled on the parade ground in their battle fatigue uniforms.

I was retired, on the spot.

For the rest of the day, we drank the beer the SF provided. We told stories, shook hands, slapped backs.

Nobody worked the rest of the day, which was how I envisioned it.

At the end of the official retirement ceremony, on February 1, 1972, I stood and watched as two hundred or so of these great men, in formation and in uniform, sang "The Ballad of the Green Beret."

As they sang, I said, "Good-bye, Special Forces," and cried without shame.

As the tears rolled down my cheeks I had no

way of knowing my future would include so many more opportunities to work with the great men of Special Forces.

In many ways, this ending—this sad, unexpected ending—was nothing more than a beginning.

CHAPTER 6

Out of combat, out of the military, out of commission. That's how I felt in the years following my 1972 retirement from Special Forces. I found myself back in Texas in 1975, needing to make some money, so I went to work for the U.S. Postal Service. It seemed like worthy employment, and within a few months I was as bored as a human can possibly be. After nearly twenty years in SF, much of it in combat, sorting mail doesn't scratch the same itch. Not even close.

I stuck with the postal work until I received a phone call on July 20, 1977. The person on the other end didn't identify himself, but he asked simply: "Billy, are you ready to travel?"

I recognized the voice as being a friend from my SF past. Of course, my answer was as simple as the question: "Roger. Just tell me when and where."

"You'll be contacted tomorrow with the details," he said.

The next day, a caller told me to arrive in a specific guest room in a northern Virginia hotel at 1500 on July 25. I was instructed to pack for one year of travel to a warm climate. "The area will be Africa," the voice said, revealing nothing more than the destination's continent.

Ideas began to form in my head. I was almost certain the CIA was forming a ground team for work, wherever and whatever it might be. I wasn't too worried about the area or its potential dangers, because I was tired of screwing around in Texas without much to do. My desire to live on the edge had not waned, and the postal service was nowhere near anyone's idea of the edge.

I met my SF contact, a man I will not name, in the proper place at the proper time. He gave me a cursory briefing.

"The work is in Libya," he said. "You will be traveling to Libya within the week and will be working for some people who will not identify themselves at this time."

This all added up to one conclusion: an agency project. I was delighted, and more than a little surprised. The agency had been decimated by the closeout in Vietnam. When Admiral Stansfield Turner took over the CIA in February 1977, he handed out pink slips to eight hundred career covert operators, rendering the outfit nearly impotent. I had aspirations of working for the agency, but Turner's ideas had me utterly convinced I would not be included in any post-Vietnam CIA. I was the exact type of man

Turner would not have: aggressive, forward-leaning, proactive, and verbal.

We processed at the Libyan Embassy for a visa, then spent a couple of nights doing the city in the D.C. area. It was a lively place in 1977, and I had some friends who were ready for a party. With months of serious and lonely work in Africa awaiting me, I contacted as many of my old comrades as possible, thereby ensuring a good time.

There was no premission training for this trip. When I inquired, the team leader told me, "Billy, you're able to do what we need, so don't worry about special training or complicated briefings."

The team consisted of four men, all former Special Forces—Luke T. was a medic; Chuck H., a communications man; and Charles T., a light-weapons man. I was considered an intel-weapons man, fair with the Morse code key. Chuck H. and I had a total of thirteen years in Vietnam combat, and Luke T. had four. Charles T., the light-weapons man, was the neophyte of the unit—three years of Vietnam combat. This was one seasoned group.

We were called to a briefing in the D.C. area that was conducted by a "lawyer" representing an ex-agency operative named Edwin Wilson. It was during this briefing that I discovered the work was not an agency job at all. Instead, it was work to be done for Wilson himself.

My bullshit antennae went on high alert upon hearing this news. I knew the CIA sometimes organized and funded Black Operations, using heavy cover to conceal its role. So, even when told the job was Wilson's private endeavor, I felt certain it had to be sponsored by the CIA.

This assignment was paramilitary in nature, training a Libyan Special Forces unit working for the boss of Libya, Colonel Muammar Qaddafi. This got my attention *tout de suite*. A chill ran through me when I heard this news. I was justifiably concerned that tackling a mission in Libya—not exactly a chum of the United States—could place us nose to nose with the U.S. State Department.

Pissing off the State Department was not recommended, but I remained involved. My intuition continued to tell me the agency's hand was guiding this operation.

To cover myself, I attempted to renew old contacts at the CIA in an effort to ferret out information on Ed Wilson and the Libya connection. Unfortunately, the old friends who had not been cut down by Turner's ax—there were many—professed no knowledge of Wilson's current interests. Wilson himself never represented himself to me as a current CIA executive.

This supposed lawyer repeatedly asserted that all was aboveboard. We would be training the Libyan Special Forces in the basics of infantry tactics. He shrugged. There was no legal risk, he said. No controversy.

To be frank, I wanted to be involved. I'd had enough of the dull post office work. I longed to be back on the edge of something covert and dangerous. This may have helped to convince me this thing was an agency-initiated covert operation, very black in cover. Therefore my vote was *"Yallah Emshee."* (Arabic for "Let's go.")

The phone in my hotel room rang late on the night before our departure from Dulles Airport. A male voice—unidentified and unknown to me—began by relating some names of Special Forces personnel he knew I would remember from Vietnam. With those "credentials" established, he asked me to meet him at a small restaurant in Arlington for some important information regarding my planned trip.

By this time, my concern had advanced to alarm. I was somewhat confused by the many moving parts in this operation.

I told the caller, "Roger, I'll be there," and promptly drove to the restaurant, where I met a man who said his name was Pat. I did not know him. By way of introduction, he showed me his CIA identification card. I studied it carefully and eventually nodded my head. It was legitimate.

"I know about your trip to Libya," Pat said. "It is not an agency-backed operation."

I nodded, still not sure what to think. Would the cover be this deep? Would the operation be this "Black"?

I chose to let Pat do the talking. He was very matter-of-fact.

"Billy, you can do yourself and others a great favor by taking some photos while you're in Libya. Do not divulge our meeting to anyone else on the team, or Ed Wilson. Take photos of

Libyan officers, foreigners, missile sites, and other things that might be of importance to us."

Pat opened his briefcase and handed me a Pentax 35 mm camera and several rolls of black-and-white film. I was given a Virginia telephone-contact number and asked to call the number upon return to the United States. I was also given a contact recognition phrase-and-reply to be passed at any future meeting or telephone contact.

I knew damned well this was the way the CIA did business—this was classic Agency Trade-craft—so I made up my mind to accept the offer. Pat closed our meeting by saying, "If the photos are worthwhile, you will receive compensation."

He waited for my answer. The mixed signals and unusual events of the past few days swirled in my mind. I began to see this photo assignment as a small slice of sanity amid the uncertainty of the operation. In effect, it gave me an ace in the hole, some protection in case questions were raised at the State Department level.

Finally, I looked at Pat and said, "OK, let's do it."

Our trip to Libya included a stopover in Geneva, Switzerland, where the four of us met Edwin Wilson at the airport. He gave the team a very cursory briefing and told us we would be flying to Tripoli, Libya, that day. Once there, we would meet a guide named Mohammid Fatah (alias).

We were met by Mohammid at the airport in Tripoli. He zipped us through Libyan Customs

A young Special Forces master sergeant, 1964.

RIGHT The pure jungle: Vietnam, 1963.

BELOW At Walter Reed General Hospital, October 1965, with Colonel/Dr. Arthur Metz (LEFT) and Brigadier General A. E. Milloy. I was a patient in the lower-extremity amputee ward at the time, and here I was being presented with the Silver Star for Gallantry and the Commendation Award for Valor.

Receiving the Legion of Merit from Colonel Elmer Monger (now deceased), Fort Bragg, North Carolina, 1968.

Here I am sharing a moment with one of the Cambodian troops who worked with me in Ba Kev, Cambodia, during our sixty-two-day SOG operation 32 kilometers behind enemy lines.

Two hours prior to HALO insertion onto the Ho Chi Minh Trail on June 21, 1971, I'm flanked by Master Sergeant James D. Bath (LEFT) and Staff Sergeant Jesse Campbell (RIGHT). Exiting a C-130 at 20,000 feet in the rain, headed for the enemies' home, produces a tight sphincter, to be sure, but away we went.

LEFT This shot was taken in the Libyan Sahara in 1979. On my right is my friend Mohammid Ageby.

BELOW I'm sitting atop a water well 250 kilometers south of Benghazi, Libya, in 1979, as a flock of camels drinks from the well. These animals are highly valued, worth $50,000 each.

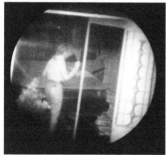 

OPPOSITE Carlos the Jackal by his Toyota Cressida outside his apartment in Khartoum, Sudan, August 2, 1994. These photographs were shot with an 800-millimeter lens from 130 meters south and 21.5 meters above the apartment building. And no, he is not pointing at me. Carlos had no idea at the time, but his final days of freedom were winding down.

TOP Carlos, the famous ladies' man, in the doorway of his apartment, August 6, 1994.

ABOVE These shots were taken on August 14, 1994, with an 800-millimeter lens attached to a Canon camera with a Litton night lens inserted: LEFT Carlos the Jackal's wife, Lana ASJ, returning to the apartment to pick up items following her husband's arrest and departure from the Sudan. RIGHT A Sudanese official, believed to be Vice President Hassan al-Turabi, arrived to assist Lana ASJ.

TOP On the streets of Gardez, Afghanistan, with local police, January 2002.

BOTTOM The barrenness of Afghanistan is on full display in Lowgar Province. Here I am, at an elevation of 7,500 feet, with the bad guys on the higher ground, 2001.

PDQ and took us to a beach hotel outside Tripoli on the Mediterranean. The hotel was adequate. Each floor had a veiled Arab female—a watcher—who kept tabs on the comings and goings of each person who resided on the floor.

At this time—August of 1977—Libya was nose to nose with Egypt in a border dispute, and I was beginning to get the impression we were being enlisted to help the Libyans achieve their objectives militarily. Our team was called to meet the Libyan intelligence minister, Abdulla Hajazi. When Hajazi walked into the room, I looked into his cold black eyes and thought to myself, "Now here is one mean son of a bitch." He spoke to us in Arabic, with Mohammid translating. (I attended a twelve-week Arabic crash course in Monterey, California, way back in 1957. Twenty years later, my Arabic was almost gone, but I knew it would return as I continued to interact with the Arabs in Tripoli.)

"Are any of you demolitions experts?" Hajazi asked.

Luke T., the medic, said he had completed the Special Forces demolitions training. I had also but did not consider myself an expert, not by a long shot.

"Have any of you trained soldiers before?" was Hajazi's next question.

Each of us answered that one in the affirmative. As a team sergeant in Special Forces and an intel sergeant major, I had planned clandestine Unconventional Warfare (UW) training in several countries, including the Philippines, Viet-

nam, Laos, and Cambodia. I could prepare a
training plan and teach any group of men all as-
pects of UW.

Three of us were moved to Benghazi, on the
Mediterranean coast about one thousand kilometers east of Tripoli. (Chuck H. was to travel for
procurement; he was not to be part of the training team.) We were housed in the Omar
Khayamm Hotel in downtown Benghazi. A nice
place, with adequate rooms and the omnipresent
female "watcher" on each floor. Each time one
of us would depart the room and pass the
watcher, she would turn to her notebook and
write down the time and the room number of the
person passing. Whatever else she wrote was left
to our imagination.

In Benghazi, commando captain Abdu Salem
Hassie, a native Libyan who was fluent in Russian and English, talked to me about teaching basic weaponry and infantry tactics to the
commandos stationed at the base in Benghazi.
The Libyans were extremely interested in recon
team fundamentals such as ambush, raids, demolitions, aerial insert, and HALO.

There was a question as to whether these men
could complete these involved and daring tasks,
but I took this job seriously regardless. I prepared a four-week training schedule to teach
these tactics to about twenty-one commandos.
Team members Felipe and Henry would assist
with the teaching.

One thing was clear from the beginning: The
caliber of these allegedly elite troops was ex-

tremely low. This didn't stop us from becoming friends with them, though—the Stockholm syndrome, so prevalent during the sixty-two days in Ba Kev, was once again proving its psychological validity.

I became close with a man I will call Captain Mohammid Ageby. He was an Egyptian, trained by the British. He was honest, and I learned to trust him. Abdu Salem was the political officer of the group; Mohammid Ageby was the leader of the commandos.

The more I became trusted by the Libyans, the easier it became for me to begin taking photographs for the agency. At the beginning, they did not allow me to take photos, period. As time wore on, they paid no attention to me as I took dozens of photos. I was not an expert, not by a long shot, but my Special Forces training in taking clandestine photos made me competent with a 35 mm camera. The last time I had taken and developed photos with any regularity was during SOG, when our recon teams took hundreds of photos.

For me, this mission had two fronts. I was training the Libyan commandos, sure, but I was always keeping one eye peeled toward potential photo targets. In the conduct of our training, I requested helicopter support to move our teams by chopper to the target areas. A U.S.-made CH-46 helicopter, flown by a pretty fair Pakistani pilot, was assigned to the commando section. To board the chopper, we took our teams to Benina International Airport in Benghazi.

As badly as I wanted to take photos of the airport and the Russian MIG aircraft on the tarmac, I knew better. In time, I suspected I would be able to accomplish this task, but I was not interested in raising suspicion during the early stages of my stay in Libya.

After loading the CH-46, our route took us directly over Russian SAM sites. I took my time, but after two months I discovered I was able to photograph these sites. I snapped the shots casually, as if the photos were of little importance. As our helicopter flew directly over the SAM sites, I was able to take very fine shots of the SAM missile positions within those guarded sites. I discovered something that would prove to be invaluable: These commandos loved to be the subject of photos, so I proceeded to use this to my advantage, posing them with important sites in the background. This tactic worked especially well as I photographed the early warning sites located in the Green Mountains east of Benghazi.

Our training of these Libyan commandos was done in all sincerity. Felipe A., Henry C., Charles T., and I worked hard and conscientiously to teach them the techniques their leaders considered important. In time, however, all doubt was removed from my mind: These men simply were not commando material. They could jog pretty well, and they could static-line parachute jump adequately. Beyond that, there wasn't much I could say about their skills.

This was my first experience in an Arab country—the first of many, as it turned out. I spent a lot of time observing and analyzing the customs

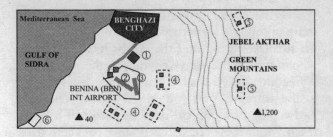

① COMMANDO CAMP
② BENINA AIRPORT
③ MILITARY AIRPORT
④ SAM SITES (RUSSIAN)
⑤ BENGHAZI PROTECTIVE SHIELDS
⑥ COMMANDO TRAINING AREA

DISTANCE FROM BENGHAZI TO
BENINA AIRPORT — 19 KM

and attitudes, curious as to what made these people act the way they did. As time wore on, I became convinced there were cultural truths at work here, truths that held for the military in Libya, Sudan, Egypt, Jordan, Yemen, Iraq, and Afghanistan.

First, it is my experience that Arab units lack the middle-management necessary to run successful military operations. This middle management—the noncommissioned officers—is the backbone of the U.S. military. The Libyans had no NCOs with authority to make field decisions. Arab units must receive orders from the topmost officer before a decision can be made. We saw this most recently in Afghanistan and Iraq, where the entire war effort collapsed when the main command structure was compromised. It is no surprise to me that the Arabs have lost every war in which they have participated.

Second, the Arabs have a terrible penchant for losing interest, pretty damned quick. The care and repair of technical devices does not interest them, and they consistently fail to keep their items fit for battle. As I taught the Libyans, I could see their eyes glaze over when I discussed map reading, the use of compasses, and communications equipment. They have no interest at the ground level because they don't believe they need to understand the information. They have always—and I mean *always*—depended on their leaders to handle technical devices.

Third, the Islamic religion is far and away the most important aspect of every Arab's life. We are gaining a better understanding of this truth, but it still takes awhile to sink in.

Religion, quite frankly, can get in the way of training and practicing commando operations. In Laos in 1961, while working with a Laotian battalion of infantry, I watched the soldiers place a small Buddha between their teeth while entering battle. Then, once in the battle, these same soldiers fired at the enemy with their eyes closed. This, as you might expect, resulted in many Laotians losing their lives.

The opposition, the Pathet Lao, did not enter battle with Buddha icons in their mouths. They also shot with their eyes open. Buddha, in this case, could not stop the bullets of the enemy. The well-aimed bullets killed their praying asses, to be perfectly blunt.

I have seen Arabs enter battle truly believing they will prevail, that Allah will provide guidance on the battlefield. Why listen to NCOs when Allah will guide you? I am not belittling Allah or Buddha or the Christian God, but I am saying battles are won by soldiers, NCOs, and officers who understand the basics of combat—fire and movement, followed by departure once the target has been neutralized. If plans are sensible and orders are followed, religion has no bearing on the outcome of battle.

My quick assessment of the Libyan commandos was blunt and unsparing. They were decent young men, but they were not fit for combat. They would not match up in battle against any other army with which I worked.

I remained with the commandos until January 1978, then made a two-week trip back to the United States. I handed off twenty rolls of film to my contact at Dulles Airport.

After two weeks, I returned to Benghazi, this time alone. In August 1978, when our year was completed, my Special Forces mates decided not to return to Libya. I wished to return, however, mostly because Abdu Salem and my friend Mohammid Ageby asked me to teach UW tactics to a group of men in a secret base in the Green Mountains.

This base intrigued me. The Libyans were extremely guarded and hush-hush when they spoke of this base, so I knew it had to be important.

I agreed. Once again, my mind was filled with competing motives.

I wanted to help my friends.

I wanted the challenge of teaching the Libyans UW.

And I wanted to get photos of that base.

My life took an unexpected turn in September 1978 when I returned to the United States and met a very fine lady in Austin, Texas. Karin Zull was a German linguist who taught on a part-time basis at the University of Texas. She was introduced to me by a lawyer friend of mine, and we hit it off immediately. She was extremely bright and had a wonderful attitude about life. It didn't hurt that she understood, appreciated, and even shared my sense of adventure.

But when it came time to depart for Libya once again, one month later, I was reluctant to leave. I delayed my departure long enough for Karin and I to be married in Hawaii in January 1979. We moved to Oahu, where Karin would eventually work at the Pearl Harbor Naval Base.

In February, a month into our marriage, I received a call in Hawaii from Libya asking me to return the next month to continue guiding and training the Libyan commandos. Karin agreed to move from our home in Hawaii to Frankfurt, Germany, where she would wait for me and live with her sister.

From Frankfurt I traveled back to Benghazi and met with the two commando officers I had worked with, Mohammid Ageby and Abdu

Salem. I expected to return to training the country's inferior commandos, but the two officers had a surprise for me.

They asked me to travel to Tobruk to meet with their boss, Colonel Fatah. When I asked for an explanation, the captains were uncharacteristically evasive. I agreed nonetheless, and we drove two hundred fifty kilometers to Tobruk, east of Benghazi.

In a country where the leader—Qaddafi—is a colonel, Colonel Fatah was a big-time and important leader. A meeting with him was not an inconsequential event, and I had been given no indication why he requested my presence.

Colonel Fatah wasted no time getting down to business.

"We want you to cross the Egyptian border and photograph positions in Egypt," he said.

I paused, considering the implications of this request.

Before I could respond, Colonel Fatah put it more bluntly: "We need you to spy against Egypt for us. We will pay you well."

They wanted me to be part of their war effort on the Egypt-Libyan border. They wanted me to perform Special Forces duty for them—cross enemy lines and provide intelligence that could aid their cause.

"No, no, no," I said quickly, attempting to distance myself from any sort of spy mission for an enemy of the United States. I told Fatah, "I'll be a sitting duck for a criminal charge from the United States."

Fatah looked at me quizzically. He did not see how I could get into trouble.

I saw very clearly how I could get into trouble. Under no circumstances would I do this. I told Fatah to forget it.

The Libyans did not expect me to refuse this assignment, and they thought a little less of me because of it. I didn't care, though, not at all. The idea of crossing into Egypt and spying for Qaddafi was completely out of the question. Had I been able to meet with an agency officer and explain the request, the CIA may have asked me to complete this sort of work. No doubt it would have enabled me to become more trusted by the Libyan officials. It would have allowed me increased access to secret areas. The benefits were undeniable, as was the danger. However, this possibility did not present itself, and Fatah needed an answer on the spot.

Following the meeting, Mohammid Ageby and I moved to the secret base of Darnah in Jebel Akthar (Green Mountains), approximately a hundred kilometers east of Benghazi. During the winter and early spring months, North Africa was one cold area. The mountains, of course, were bitterly cold, and the vast desert area we used for training was chilled by winds from the Alps that whipped unabated across the Mediterranean toward the coastal dunes near Benghazi. The landscape was spectacular and harsh, almost taunting in its severity.

I assessed the men I would be training in Special Operations–type warfare, and I knew for a fact they would never be able to conduct commandolike operations against an enemy. Completing an operation, in my estimation, was totally out of the question. Nonetheless, I began by teaching them infiltration by Zodiac watercraft. I also continued taking photographs with my trusted Pentax camera. Targets in the Darnah were many, and I stayed on the alert for potential photo targets.

I knew the commandos would be uneasy about the prospect of learning water infiltration tactics. During one of my previous training sessions in the lowland Benghazi area, I found that just two—two!—of the twenty-one men in training could swim. Now, these men were training to be commandos, for God's sake, and they couldn't swim. The Libyans, for whatever reason, did not believe a commando needed to be able to swim anywhere.

(Accidents were not uncommon during our training. Earlier, during a training session in the Mediterranean, tragedy struck. One of our trainees fell overboard and drowned. Thankfully, I was not in the area at the time, so I was not blamed.)

I began to sense a growing hostility from some of the commandos and Libyan officers. Some of it was attributable to my refusal of Colonel Fatah's request to spy on the Egyptians. In addition, the incompetence of the commandos was eroding some of the goodwill I had established. These men, unfortunately, simply could not be trained to complete the operations I was accustomed to completing. They were often unresponsive to a frustrating degree. In fact, I actually punched one or two of the *jundis* in the stomach for not obeying their officer, and this loss of face didn't sit well with these commandos. I have to say, it sure did work at the moment, especially when the orders needed to be carried out immediately in live-fire drills. Failure to obey sometimes placed the entire unit in jeopardy, and a good punch in the stomach got the attention of the person being punched. I noticed they seemed to listen to—and obey—orders more rapidly following a good sock in the gut.

We continued our training at the secret base in Darnah through the summer of 1979. I took as many photographs as I could without raising suspicions. I was near the MIG-25 parking area and came in contact with many Russians who piloted aircraft for the Libyan Air Force. I got many excellent and strategically important photos from this location.

However, morale waned and hostility grew. The hostility was most evident from Abdu Salem. His resentment came across loud and clear, so I asked my friend Mohammid Ageby about it. He reluctantly told me the resentment against Americans was building among the officers.

"Billy," he said, "they've discussed throwing you out of the country."

Our conversation prepared me for the advice Mohammid gave me during the first part of November.

"Billy, you must think about getting out of the country. The commander [Colonel Qaddafi] is furious with the Americans."

On November 4, 1979, word reached us that a mob of three thousand people described as students furious at the United States for providing asylum to the shah stormed the U.S. embassy in Tehran and took fifty-three U.S. personnel hostage. This, of course, went on to become a lengthy and embarrassing siege on the American psyche.

That day, Mohammid Ageby came to me in a frantic rush.

"Billy, you must leave tonight. You'll will be thrown in jail if you don't get out of the country."

The Libyans were furious with America for this perceived betrayal of Islam, and they showed their displeasure by burning and sacking the U.S. embassy building in Tripoli. This was not a good time to be an American in Libya.

Mohammid had arranged a flight for me to Frankfurt. It left in two hours. I wasn't sure I could make it.

"You can't take anything, not even money," Mohammid said.

He handed me the ticket to Frankfurt. Since Karin was in Frankfurt, I called her and asked her to meet me at the Frankfurt International Airport. I had a credit card and no money. I rushed to the airport in Benghazi and shook hands with my friend Mohammid Ageby. It was a short good-bye, and I was sad to leave another friend in a bad spot.

I met Karin in Frankfurt with only the clothes on my body. We returned to Hawaii via Anchorage, and another chapter of my life came to a close. I was able to call my Virginia contact number, and the man on that end set me up with a contact in Hawaii who accepted my dozen or so rolls of photos and forwarded them to the CIA.

I later heard from reliable sources that my friend Mohammid Ageby—a man who may have saved my life—was executed for his role in an attempted coup against Qaddafi. Damn, my good friends seem to end up dead all too often.

My Libyan experience was a pivotal event in my life, which is remarkable considering the maze of motives and countermotives of those who were involved either directly or indirectly. Edwin Wilson, the ex-CIA agent who organized the work, made millions of dollars hiring intelligence and military people to work for him across the globe. In 1976, federal agencies began investigating his activities. Wilson was accused of "false-flag" recruiting, with several witnesses testifying that Wilson misrepresented himself as a current CIA executive. (Again, Wilson never misrepresented himself to me.) The five-year federal investigation into Wilson's dealings led to his arrest on several charges, including one bombshell: Wilson had allegedly sold more than twenty tons of C-4—a plastic explosive ideal for terror operations—in 1977, the year I was recruited to train the commandos. Wilson was eventually convicted on three counts—the sale of the C-4, attempted murder, and the illegal export of arms—and sentenced to fifty-three years in prison.

Wilson served twenty years before his conviction for selling the C-4 was overturned on October 29, 2003, by a federal district court judge in Houston. In overturning the conviction, the judge ruled that prosecutors used false testimony to undercut Wilson's defense. In his appeal, Wilson was able to produce at least forty examples of his work for the CIA following his retirement. Even though nothing indicated the CIA approved the sale of the explosives, Wilson was able to document the CIA's knowledge of his presence in the country. The agency had also requested his help—as it did mine—in gathering information pertaining to Libya.

Wilson's arrest and conviction made me sit up straight and take notice, but I was never called on the carpet to answer for my work in Libya.

As I had hoped, my photography work for the CIA worked as a shield against possible repercussions. And I believe the hours I spent photographing strategic spots proved to be more than simply busywork. On April 14, 1986, twenty-four U.S. Air Force F-111-Fs took off from a base in England, headed for Libya. Before it was over, Operation El Dorado Canyon dropped more than sixty tons of laser-guided bombs on five targets.

As I read the reports of the missions, one of the targets jumped off the page at me: the Benina airfield and barracks. The bombs destroyed four

Russian-made Libyan MIG-23s, five transport planes, and two Russian-made helicopters on the base I had photographed clandestinely for the better part of two years. All of the secret SAM sites, photographed so carefully, were also destroyed before the commencement of the F-111-Fs' bombing runs. I would like to think my undercover work played a role in the choice of targets for the operation. At the very least, my camera work in Libya sparked a passion for photography that served me exceedingly well in my later endeavors.

In retrospect, I never could have planned a mission that would set me up so perfectly for future work. It ignited a wanderlust that seemed to attach itself to my DNA. It was my initial exposure to an Arab country as well, and the knowledge I acquired and the observations I made would serve me well as I embarked on counterterrorist work throughout the Middle East.

Last but not least, my experience in Libya put me back in touch with the CIA and led to my work as an independent contractor for the agency. This work would take me all over the world and renew my relationship with Special Forces.

What began as a lark, something to get me out of the mailroom, became a shrewd and prescient career move. Put simply, Libya sent me right back to the edge, exactly where I belonged. As I consider my two years in Libya in relation to all that followed, one word springs to mind:

Fate.

CHAPTER 7

So now I find myself in Hawaii, without a lot of money, and the Libya gig is blown to hell. Our fifty-three hostages remain in Tehran, awaiting rescue, while the Carter administration wrings its hands and twiddles its thumbs, wondering how it got into this situation and how to get out of it. In the United States, the Army and the CIA's paramilitary unit are stagnant, stuck in the same do-nothing malaise as the government. Both organizations are being run by Special Forces haters, and it is widely known that retired SF men—decisive men of action who often had their own ideas about how operations should be run—are not welcome in the early 1980s CIA. Unless the tenor of the times changes drastically, I see no chance of pursuing my goal of becoming an independent contractor with the agency.

The fruits of my labors in Libya—the fruits of fate, as I believed—are on hold. It seems the entire military is on hold in the 1980s, and the CIA follows suit. Retention rates are down; applications for the military academies are down; morale is way down. The Army, the CIA, and our entire country are in a transitional phase, and as Karin and I live our lives in Hawaii, I realize I am in transition as well.

I'd already seen the other side of life, the civilian side, and if it looks anything like the U.S. Postal Service, I know it isn't for me. My beliefs never change, though, and my beliefs tell me my country will always need men who are willing to go anywhere and do anything to seek and fix enemies of the United States.

I know they'll get back to me. I just don't know when.

It is my opinion the post-Vietnam era turned the United States into a second-rate nation militarily. We were unable to complete even one successful operation against our enemies. We slinked away from Vietnam with a president who was forced to resign, replaced by an unelected vice president who found himself in this position only after the elected VP was removed from office. These were bad times, and it affected all branches of the service, not to mention the CIA, FBI, and every other government organization. Damn, those were terrible years for our nation.

Then we elected a very bright peanut farmer who wished to please the entire world by being nice. That's one approach, but it doesn't work. There was no commitment to the military from the top during these dog days, and the outlook

appeared bleak. There was a widespread
anti–Special Forces sentiment among the highest
brass of the military and the government in the
1980s. This was an old battle for Special Forces,
a group that had grown accustomed to justifying
its existence. The military clearly needed to
change with the times, but reaching an agreement on the type of change was not a smooth
process. Conventional generals simply didn't see
the need to develop Special Operations when the
tried-and-true methods were still available.

At the same time, the CIA's worth was under
constant attack from the liberal do-gooders who
felt the agency's paramilitary units were free-
lance renegades who needed to be eliminated.
This was a buildup that began with the secret
wars in Laos in the late 1950s and accelerated
with operations such as SOG and the overthrow
of the Allende government in Chile. The famous
findings of the Church Committee set the tone
for the times: The CIA must be neutered, turned
into an outfit littered with pencil-pushers and
paper-filers. It would concentrate on *watching*
enemies instead of *fixing* them, but it wouldn't
hire the types of people who knew how to do either.

One of the major events that altered the landscape for the military and the paramilitary units
of the CIA was the disastrous Desert One, the
failed April 1980 attempt to free the hostages in
Iran. The operation was headed by Colonel
Charlie Beckwith, who led Delta Project at the
time. From the day the hostages were taken, Col-

onel Beckwith studied the Tehran target area diligently from televised news reports. He formulated an elaborate rescue operation and eventually convinced Jimmy Carter—despite significant opposition within the cabinet—that his dangerous and risky plan could work.

And it might have worked if it hadn't been for the meddling of the Joint Chiefs of Staff, who decreed the rescue should be an all-service operation. One of the decisions, made by the U.S. Air Force commander of the rescue, forced pilots from another service to fly the RH-53D Sea Stallion choppers before these pilots were qualified for either night flying or flying through a *haboob*, or dust storm.

Desert One was the designated staging and refueling area, two hundred miles from the rescue target in Tehran. The plan called for eight Sea Stallion helicopters with Special Operations personnel on board to land at Desert to be refueled by two C-130s. Beckwith's plan called for a minimum of six helicopters to pull off the rescue operation.

The plan went to hell in a hurry. A bus, headlights blazing through the night, arrived at Desert One carrying forty-four Iranian passengers into this allegedly remote area. Delta members stopped the bus and took the passengers hostage, and shortly afterward a gasoline tanker rolled into the location. It was blown up and the explosion brightened the night sky. Furthermore, one of the helicopters en route to Desert One was forced to land after it lost rotor blade pressure. Another lost its gyroscope in the inevitable *haboob* and turned back. Of the six that made it to the staging area, one was found to be without its hydraulic pump and therefore unable to take off.

This left Beckwith with five choppers for a six-chopper operation. On the ground at Desert One, the colonel bitterly scrapped his elaborate and well-conceived plan. It simply couldn't proceed with the available resources, and like any good commander Beckwith recognized this fact and refused to send his men into an untenable situation. Then, to add further tragedy to the mix, one of the Sea Stallion helicopters collided with one of the C-130 transports as both were attempting to depart the area. The Sea Stallion exploded, killing eight men aboard.

The world never learned whether this daring rescue, code name Eagle Claw, would have been successful. There is no doubt, however, that its failure had wide-reaching repercussions. In a broad sense, it was the first U.S. confrontation with modern Islam, the first time U.S. soldiers died while preparing to do battle with this relatively new and burgeoning threat. Also, in another consequence that impacted my life, this was a botched Special Ops mission in a time of skepticism regarding the worth and necessity of Special Operations. The timing couldn't have been worse.

In his August 1, 2003, arrival message upon being installed as the thirty-fifth chief of staff of the U.S. Army, General Peter J. Schoomaker said:

Twenty-three years ago I stood in another place—in the Iranian desert on a moonlit night at a place called Desert One. I keep a photo of the carnage that night to remind me that we should never confuse enthusiasm with capability. Eight of my comrades lost their lives. Those of us who survived knew grief . . . we knew failure . . . but we committed ourselves to a different future.

There were some important things we did not know about the future that night. We did not recognize that this was a watershed event . . . that the military services would begin a great period of renewal that continues to this day. We did not know that we were at the start of an unprecedented movement to jointness in every aspect of our military culture, structure and operations . . . a movement that must continue. We also did not realize that we were in one of the opening engagements of this country's long struggle against terrorism . . . a struggle that would reach our homeland and become known as the Global War on Terror.

Men who worked in Special Operations have always fought against the tide in the U.S. military, so the stagnation of the early 1980s was simply an old messenger in a new suit. In the years immediately following World War II, conventional generals did not want any type of Special Force to be created. These old bosses fought annually for their money allocation from Congress, and they weren't interested in sharing the wealth with any up-and-coming operators they perceived as uncivilized rogues. The admirals in their dress whites and tassels would strut and moan that without new battleships, the United States was doomed. The Air Force repeated the refrain, stressing the need for the latest aircraft. (We now realize they had a right to this argument.) The Army, meanwhile, plodded along with its tanks and infantry. The conventionalists believed (and with some credence at the time) that warfare should be a battlefield slugout, toe-to-toe, both sides blasting away with tanks and artillery, with sufficient time for the licking of wounds in between.

These honorable battle commanders loved these full-frontal confrontations; they gleefully anticipated the moment when they would send forth their main line of resistance (MLR to old military men) to plod, slug, and inch their way forward, killing everything and most everyone in its path. The U.S. Air Force (as it became known in 1947) was used strictly for strategic targets, usually bombing dozens of miles behind enemy lines. Close Air Support (CAS) bombing within a few hundred meters of friendlies in some instances and within fifty meters of friendlies in the case of SOG, was not part of the plan.

Frontal-assault, slug-it-out battle was considered the honorable way to conduct warfare. In his papers, General Patton made it patently clear that this venerable and highly principled form of war was viewed as noble by those conducting the war. Patton, the great legend, was not much concerned with friendly casualties; he was interested in employing the full forces of artillery and infantry to crush enemy forces on the battlefront. There wasn't a whole lot of room for sentiment in that equation.

The Patton formula was repeated over and over, with victory in battle after battle eventually adding up to a victory in war. This is fine, or was fine, but it quickly became dated as technology accelerated. Conventional warfare demanded too many troops, tanks, battleships, and dumb-bomb aircraft to win a war. Too many lives are lost, too many days wasted. *Clank-clank I'm a tank* was no longer the best use of military might.

When Special Forces came along in 1952, consisting of a few hundred NCOs and officers belonging to the U.S. Army, an idea began to hatch in the brightest military minds. These SF men, trained to the *n*th degree, with specialties beyond the scope of all others, could deploy small teams into enemy territory and—if supplied properly—fuck with the enemy to the max from within. These Special Forces men had to be so dedicated and well trained physically and mentally to overcome the fact that friendly support was often hundreds of miles away. This logistical nightmare did not deter their determination to disorganize and defeat *any* enemy. Our Special Forces leaders visualized the change in ground warfare as a broad manifestation of these concepts—planning, training, secrecy, ambush, shock, and resolve—along with the all-important smart bombs used by the U.S. Air Force.

This did not sit well with the tank commanders who were certain they had won World War II, or the artillery brass who knew their high-angled fire had defeated the Axis. These generals wanted their money, and this frigging new outfit—filled with snake-eaters and scum—was going to take it away from them. So this battle raged, and in 1956 the anti-SF folks won a partial victory when Special Forces was reduced from twenty teams to twelve.

But then Special Forces began to train foreign personnel by living among them. SF men learned the languages—albeit basic military jargon in most cases—then trained, lived, and died with these indigenous personnel. This was a great success, and it caught the attention of the CIA, which liked the idea of small units infiltrating into the enemy's lair. In 1958, the agency asked for SF teams to move in with the battalions of infantry of the Laotian army. This controversial mission—Operation White Star—marked the beginning of the ongoing marriage between Special Forces and the CIA. I went into Laos as part of White Star in 1961, toward the end of the operation, and was one of the first Special Forces men to enter Vietnam in 1962.

Since the CIA was picking up the tab as part of its paramilitary unit—then called Combined Studies Division (CSD), now dubbed Ground Branch—the conventional generals didn't find these developments too distressing. But that changed with Vietnam and the creation of the 3rd, 5th, 6th, 7th, and 8th Special Forces groups. This cost bucks, and the infantry generals found their best men, quick to realize the potential of the new concepts, were volunteering for Special Forces training. Money drained from the Army's bit of the annual fund, so the chief of staff and his generals began to piss and moan all over again.

These old conventional generals had—and some still have—a difficult time understanding the dynamics of Special Operations. They don't see how the job can be completed by a few men a few hundred miles deep in enemy territory, with a world of technology at their disposal. These men are placed through unconventional means by fine aircraft into targets, where they own the high ground and call in brilliant bombs that destroy the enemy in his factories and training areas. To those of us who have seen both conventional and unconventional combat, it is not difficult to choose sides.

The conventional brass see warfare as crunching the enemy in the tracks of their tanks, taking the ground—holding the ground as theirs, raising the flag, and continuing the march forward. Well, nice people, you can see what holding the ground did to us after the end of the official hostilities in Iraq. That country is an example of what a "hold territory" war causes. Holding and protecting the ground allows the enemy to deploy small groups against our soldiers. Of course the old conventional brass doesn't give a good shit about this, for they will simply withdraw to their enclaves when the order arrives, then go about the business of drawing their pay while they study the historical tactics of wars past.

As always, there were political issues to consider when analyzing this new opposition to Special Operations. As soon as these conventional four-star types admitted the superiority of this unconventional but great form of warfare, they were obsolete. They were gone, finished, put out to pasture. The money for their conventional battle gear and machines would be taken away. They would be forced to conform or be finished, so these four stars fought to their deaths trying to convince Congress that these Special Operations groups (Special Forces, the SEALS, Rangers, the U.S. Air Force Special Tactics Squadron, as well as the Combat Talons and other Air Force Special Operations personnel, and the Force Marines) were a waste of time, energy, and money. However, with Special Operations proving itself time and again, these old boys have had to retreat to their respective corners, tending to their wounds and settling for a smaller take of the loot.

This brief history lesson is pertinent to the situation in the early 1980s and pertinent to my situation as I struggled to plot a new course for my life as I entered my fifties. With Carter out of office, President Reagan made upgrading the military a priority in the first four years in office. He spent a lot of money on the conventional military, especially the Navy, but along with it he

oversaw the creation of the U.S. Joint Special
Operation Command (JSOC). The brightest of
the Special Forces officers foresaw great things
from a combined force of U.S. Air Force, Special
Forces, SEALs, Rangers, Force Marines, and
Delta Force.

The advent of JSOC, perhaps unintentionally,
set in motion what is called a force multiplier.
And in this case, it was a force multiplier
supreme. A force multiplier occurs when a factor
in an equation accelerates the combined sum of
its parts exponentially. As we have seen in the
lightning-fast rise in Special Operations, technology was the force multiplier in this equation.
Three Special Forces men armed with guiding
devices to deliver smart weapons from Air Force
jets or unmanned aerial vehicles overhead onto
enemy targets three hundred to eight hundred
miles deep in the enemy's territory—wow, that's
a force multiplier the likes of which the military
world has never seen. Each component has
worth in its own right, but together they produce
something astronomically more powerful.

I relate the military transition of the 1980s—
from a strict reliance on the conventional to an
embracing of the technology-based unconventional—to a dentist telling a forty-year-old he
has to lose his trusty old molars. "But why?" the
forty-year-old might ask. "They've served me
well to this point." To which the dentist would
reply, "They're no longer efficient, and besides,
we're engineering food you never have to chew."

The proof, of course, came in Panama, Desert

Storm, Afghanistan, and Iraq. This new brand of warfare proved its worth repeatedly, and even the conventional generals had to concede as much. Consider the perfect example: Afghanistan. We soundly defeated that nation and its Taliban extremists in three months using small groups of Special Forces men and local anti-Taliban forces. Laser, infrared, GPS guidance, along with satellite overwatch—it was an unbeatable combination. And to think, just two decades earlier, the Afghans dealt a horrendous, demoralizing, and embarrassing defeat to the Russians and their *clank clank* conventional methods.

This transition was painful, spanning several years and many political battles. To this day I can see the little old pompous admirals, splendid in their dress whites, riding in their admiral's skiff as all the ships in port pay homage to the little man as he is piped past each moored vessel in port. This tradition and honor is not to be scorned, but let's save it for stateside-peacetime bullshit. In war, let's get into the target, above the enemy, and bomb his ass out of existence. The old-time conventional brass hats, whom I salute and honor, remind me of the old Pac-Man game. If you remember, Pac-Man always jumped up and down a few times, almost reflexively, during his demise.

Where did all of this upheaval in the military world leave me? Nowhere. I was literally at sea while our nation's military was figuratively at sea. At times it appeared to me that, despite my

optimism, my days of assisting the United States
in knocking out its enemies were over. I was no
longer a young man, and the more the transition
period dragged on, the less attractive I was going
to be to any governmental agency. As the military was forming JSOC and the Special Operations Command, I knew nothing, *nada,* about
the struggle to find the best way to regroup, re-
form, and modernize the military. It wasn't easy
for them to figure it out, and it wasn't easy for
me to wait it out.

I was in an exotic place—Hawaii—but I
couldn't get an exotic job with any service. I decided to take advantage of the lethargy in my
chosen field of work to further my education in
an attempt to make myself more attractive to the
CIA when the pendulum swung back my direction. With my wife Karin's blessing, I went to
school at night, studying variously at the University of Hawaii, Chaminade University of Hon-
olulu, and Weyland Baptist University of Hawaii.
I worked during the day as a police officer at
Pearl Harbor, a job that was neither enjoyable
nor very challenging.

I was a man in transition in a country in transition.

In 1987, a retired Special Forces colonel asked
me to become a deputy chief of police at the U.S.
Army Kwajalein Missile Range (KMR), in the
Marshall Islands, approximately twenty-five
hundred miles south southwest of Hawaii. I
agreed, and we sold our condo and moved bag

and baggage way south in the Pacific, to a place now known as the Reagan Test Site.

I was responsible for the employment of roughly one hundred U.S. security officers who were flown weekly to the various islands—about twenty-five small, atoll-like spits of land—of the Kwajalein Missile Range to ensure the security of the equipment and the safety and welfare of the local Marshallese.

Sounds pretty mundane, but there's one reason Kwajalein Missile Range is more than pristine beaches, crystal-clear water, and tropical fish. The U.S. launches missiles periodically from Vandenburg AFB in California to preselected targets throughout the KMR. After launch, these big-assed missiles exit the atmosphere as they travel SSW toward KMR. On reentry into the atmosphere, the missiles separate into multiple nuke-tipped missiles, and each missile hits a designated target in the ocean. From launch to impact, this takes about thirty minutes of airborne travel.

Since our missiles were there, and this was the early to mid-1980s, that meant the Russians were there. Hungry for the technology held inside each of those missile's nose cones, the Russkies had a constant presence in the KMR area. These old boys put some thought into this endeavor; they had an electronic listening surface vessel and a submarine in the area almost year-round. These lunkheads had visions of stealing or otherwise securing our nose cones after their impact in the range area. The frigging

Russkies were too *dummerazle* to create their own guidance systems, so they figured they might as well take the next best option: thievery.

Our job, as you might expect, was to keep the Russians from completing their mission. The Russians had good reason to want those nose cones: These missiles were right on damned target, just about every time. The individual Multi-Re-Entry-Vehicles (MREV) were launched from the huge missile that had been launched at Vandenburg, and those MREVs would hit directly on selected targets within the Kwajalein atoll. And these targets were not just in-the-vicinity targets; no, they were pinpointing targets that were often simply a buoy placed out in the atoll. These targets were being struck dead center.

Of course the Russkies viewed this wonder of technology from their electronic surface vessels and submarines. As the MREVs hit their targets, the Russians would put a rubber Zodiac boat loaded with divers into the water. The Zodiac would buzz its way to the target area and the divers would hop in, attempting to steal our nose cones. Our job was to get our own divers into the water to catch these nuts.

Most of our work was in teams, but I occasionally conducted my own "singleton" missions in an attempt to mess with the Russians. My goal, though never realized, was to grab a Russian Special Forces man off one of the boats or Zodiacs and get a more detailed idea of their intentions. After a run-of-the-mill island search, I would stay behind after the other members of the search team departed. The waters were open, so we could do little about the presence of their electronic surveillance ship and submarines that visited the waters. But when their small Zodiacs infiltrated the atoll and plied the lagoon waters, they were in U.S.-leased waters and therefore fair game. We managed to scare a few of them out of their wet suits several times, but that was as close as I came to nabbing one of them. I enjoyed the solitary work of watching and devising a plan of action to outsmart the enemy; good thing, too, because I would come to employ those talents while tackling much more important enemies in the near future.

The work in Kwajalein was a real deal, and we did our jobs well. The Russians, despite the commitment of personnel and equipment, never came away from Kwajalein with a single nose cone. We didn't come away with a single Russian, either, but on many occasions we routed them from one of the spit islands so quickly and thoroughly that they left behind all sorts of gear—rations, wet suits, and so on. Their equipment, of course, was horribly crude compared with U.S. gear.

This game we played with the Russians so far removed from the U.S. mainland was all too indicative of the cat-and-mouse Cold War shenanigans that took place all over the globe. It wasn't as exciting as SOG, or my days in Libya, but it was all I had. I was dragging my heels looking for excitement, and this is what provided the adrenaline rush during the early 1980s. You take it where you can get it.

In a serious sense, this job allowed me to see the power of technology, the potential of this force multiplier. I might have been out of the military, but the military was not yet out of me. I was not a witless bystander watching these wonderful smart missiles as they repeatedly struck their intended target. My mind, as always, was working, figuring out what this meant for the future of our warfare.

I viewed it the way a man who had been in my position in Vietnam had to view it—as an exciting alternative to sitting in the jungle attempting to signal tactical air via handheld radio or mirror.

And still I waited for the tide in the States to turn. I waited for the brightest and most practical minds to win out over the knee-jerk dogooders and bring our country's military back to its historic prominence. I knew—and recent history has borne this out—the welfare of our nation depended on us recognizing the new threats and devising a plan to deal with them. I had a few ambivalent moments, but I knew they would eventually get back to me, and in 1985 I received a call from an old friend and Special Forces team sergeant Paul P.

"You interested in returning to the States any time soon?" he asked.

"Depends on why you're asking," I answered.

Paul explained that he had a line on an agency job for me when I returned. This was the break I knew would come, the sign indicating a significant shift in the mind-set of our nation's military and paramilitary units. I wasn't ready to jump right away, but I knew I was going to take Paul up on his offer as soon as the time was right. I had a few more tasks to cross off my to-do list.

I arrived in Hawaii as a high school graduate, and two years after Paul's telephone call I returned to the United States with two B.S. degrees from Weyland Baptist University. I found I enjoyed the schoolwork immensely and no doubt appreciated it more because of my age, and I decided to pursue a master's degree in Interdisciplinary Studies (with an emphasis on Criminal Justice Administration) at Southwest Texas State University in San Marcos, Texas.

So, with a master's degree in hand, I visited Paul P. in Washington, D.C., in 1989 to interview for an independent contractor's job with the CIA. I was offered a job with a unit that originated as something of a "hit unit" organized to eliminate individuals who present significant threats to the United States.

The prospect of such work was no doubt enticing, but the original intent of the unit never came to fruition. The do-gooders—agency oversight personnel and many foot-draggers in high positions—applied their pressure, and the unit was toned down to a mere surveillance unit. In effect, it was defanged by senators who had no concept of how to gather information and how to eliminate those who have our demise as their ultimate goal. These folks, though less powerful and smaller in number than ten years previous, had no idea what those of us in the clandestine arts need to complete the job at hand. We have struggled to recover from their blind allegiance to their own bleeding hearts ever since.

Nonetheless, the pendulum had swung, much in the way I expected, and when it happened I was prepared with educational credentials to go along with my combat experience. I was still young and energetic enough to be attractive to the agency, and I was motivated by close to a decade of relative inactivity. In recounting my career with the agency, I have chosen to focus on two of my highest-profile targets—Usama bin Laden and Carlos the Jackal. Many of the other assignments and targets cannot be revealed, but there were many other successes along the way. I was often recruited to be a "singleton" because I work well alone and am afraid of nothing and nobody. As the woeful 1980s became the hopeful 1990s, my country was back in business, and I was brought along for the ride. I was a hunter, a full-time hunter, at a time when the hunting was good.

CHAPTER 8

A cloud of dust flew up behind the big white Mercedes like a contrail from a B-52 as I jogged up a street in the al-Riyadh section of Khartoum, Sudan, in February of 1992. I'd seen both before—the car and the dust—so I wasn't surprised when the Mercedes pulled up in front of the house next to me and I found myself face-to-face with the man in the driver's seat. He was driving alone, as was his habit, and in the moments before the dust overtook the car and continued its billowy flight up the road, he looked me square in the face. I returned the favor, locking on to the dark, heavy-lidded eyes of Usama bin Laden.

I always knew when bin Laden was behind the wheel of that white four-door Mercedes 300 with Sudanese plate number 0990; the man zoomed in and out of those dusty streets like greased shit. Traffic laws were pretty loose in Sudan, and this character drove like he owned the place. At that point in time, it wasn't far from the truth.

I know what he thought when he looked at me, a sixty-two-year-old American jogging down this dusty street: *What in the hell is that old American bastard doing out here?* It was the first of many times I came within ten feet and face-to-face with him over the eighteen months that I jogged past his house as he was either coming or going, so I'm guessing it didn't take him long to figure out what I was doing. He was a bright man, and seeing an old white man like me, running up the street near his house, was bound to raise suspicions. I didn't give a good goddamn about suspicions, though—I was entitled to be there, and I had a job to do.

I jogged past him and continued my run, up past his house and the squadron of huge Afghan guards standing on high alert at the perimeter of the property. Even though their boss roamed freely, these guys were ready for action. Their job was to protect UBL, and I have to say they did their jobs well. With their long beards and fierce dark eyes, these men meant business. They wore local clothing rather than the long garments common in Afghanistan. They searched constantly and vigilantly, in a manner befitting men who believed they were protecting not only their boss but their messiah. Without being too obvious, I could see and count four or five guards at ground level. I could only imagine how many others might be on the roof and in the upper stories of this large, three-storied residence. As I continued jogging and watching, I knew I was being watched, and watched closely. It's a feeling I know well, and a feeling I've learned to recognize and relish. These men, I knew, were putting the equation together in their minds: An American passing close to the residence of Usama bin Laden meant their boss was of interest to the United States. Whatever I was doing there could not be good for either them or the country in which they had chosen to reside. These sentries were tough-looking bastards, and with AK-47s slung over their shoulders, they were clearly not friendly. They said nothing but stared their hatred to me through those black, deep-set eyes. As was the case with bin Laden, I looked these Alpha Sierras (assholes, as we said over portable radios) dead-on, and they stared right back with hatred in their eyes.

Everything was reduced to the sound of my own breathing and the falling of my own footsteps in the dirt. I didn't change my pace. I wasn't concerned for my safety; I wasn't doing anything illegal, and bin Laden's boys knew the best way to ruin the good thing their boss had in Sudan was to pop an American for no good reason. I jogged this route and variations of it at least twice a week during my time in Khartoum (a place we called K-Town), and over time, my jogging and their watching became something of an awkward dance.

Between February 1991 and July 1992, I was one of the first agency operatives to be assigned to watch and photograph bin Laden. At the time bin Laden was not considered an especially high-level assignment, and Khartoum was so completely saturated with miscreants and no-good bastards that my hunting wasn't limited to this one tall Saudi exile. I would cycle in and out of Sudan every six weeks, sometimes staying as long as three months if the hunting was good. Six weeks was about the average shelf life of a street operator in Khartoum, for the Sudanese security force charged with keeping an eye on foreigners (a mostly incompetent group called the PSO) had a tendency to make life uncomfortable for those who stayed too long on the streets. Nothing was easy in Khartoum; for one thing, the government had outlawed cameras and developing supplies, fearing images of the country's degradation and destruction would be distributed to the world. People lay starving on the streets of K-Town, and many of them were picked up and transported to areas outside the city and left to die a less obvious death. This was not the image of Sudan its government wanted to portray. Photographing targets was one of my main responsibilities, however, so we circumvented Sudanese law by bringing cameras into the country in diplomatic pouches, which were exempt from inspection. Dip pouches are a loophole negotiated through the State Department, and in many cases they allowed us to do our jobs. If I had been caught with a camera on the streets of Khartoum, though, I would have immediately been labeled persona non grata. The CIA chief of station would have had no recourse;

he would have kicked me out of the country *tout de suite*.

Sudan in the early 1990s was an anything-goes cesspool of a place, a free-for-all for terrorists and rogues from throughout the Middle East. It was, and still is, embroiled in a bloody civil war between the Christian majority and Islamic minority. Sudan was the center of radical Islam, and its rulers allowed terrorists to build their camps and roam freely. They were not only welcome; they were VIPs. They lived in relative prosperity within the city while vast slums surrounding K-Town were filled with starving refugees of the civil war.

This was the home of the stateless runaway, with a ruthless son of a bitch on just about every corner. Bin Laden, whose attitudes and intentions were just beginning to attract the attention of the United States, fit in perfectly. Put it this way: I could go on a long run from my residence in the al-Riyadh section and run past the homes or support sites of bin Laden, Abu Nidal, various members of Hezbollah, the Egyptian Gama'at al-Islamiyya, the Algerian Islamic Jihad, HAMAS, the Palestinian Islamic Jihad, and a bunch of badass Iranians loyal to the Ayatollah Khomeini's "Death to America" craziness. Nidal was spending most of his time in Tunisia and Libya, but he filtered in and out of Khartoum. The Blind Sheikh, Omar Abdel Rahman, received his U.S. visa from Khartoum in 1993 and proceeded to plan the first World Trade Center bombing. All those bastards were there, praying and training to kill Israelis and Americans. What a fucking place, and by the end of 1993, you could add the most famous terrorist of them all—Carlos the Jackal—to that list.

These bad guys came to this African city at the confluence of the White Nile and Blue Nile rivers with the blessing of Sudan's vice president, Hassan al-Turabi. Al-Turabi, educated at the Sorbonne, was responsible for turning Sudan into an Islamic state in 1991, and as head of the National Islamic Front, he considered himself the spiritual leader of not only his country but the entire continent of Africa. He was once quoted by a newspaper in Sudan, *Al Rai Al A'm,* as saying, "We want to Islamize America and Arabize Africa." He allowed the terrorists who shared his beliefs to run rampant in K-Town. Of course, he had to get something out of it, either financially or, if that failed, for his country. The Iranians, through Khomeini's regime, were funneling serious amounts of money into Sudan. The Sudanese president, a figurehead general named Omar Hassan al-Bashir, was really nothing more than a puppet for al-Turabi. Al-Bashir attempted to court favor with the United States while simultaneously going along with most of al-Turabi's dictates. Every time the president raised an objection to al-Turabi's insistence on harboring these criminals, the Iranians would send al-Bashir another new Mercedes, and he'd be just fine with everything once again. In a country with a per capita annual income of roughly $330, terrorism was an economic boon.

UBL—this shorthand and the spelling "Usama" were choices the agency made at the beginning of his notoriety—was a VIP in Sudan. The story of how he found himself in K-Town is a Byzantine tale of zealotry, war, and perceived betrayal.

The bin Laden family was famous in the Middle East long before UBL became the world's most-wanted man. His father, Mohamed bin Laden, was a Yemeni who decided to leave his home country in 1930 and take a thousand-mile camel caravan to Saudi Arabia in 1930. An illiterate who spent his whole life signing his name with an X, Mohamed worked in construction, started his own company, and got his biggest break when he undercut the competition to win contracts to build palaces for the House of Saud in Riyadh. In an October 2001 investigative piece in *The Guardian* newspaper of London, a French engineer was quoted as saying Mohamed bin Laden "changed wives like you and I change cars." Mohamed was said to have three permanent wives and one rotating wife; Islamic law allows for four wives.

Usama's mother was said to be either the tenth or eleventh of Mohamed's wives, a Syrian named Hamida who wore designer clothes instead of a veil and was known as Mohamed's "slave wife." UBL was eleven and already accustomed to the good life when his father died. The son developed his obsession with Islam in his late teens, but that obsession battled with family duty to study engineering and contribute to the family's

construction business. While studying for his
university education, which he completed in
Saudi Arabia rather than overseas like most of
his siblings, he combined the two. His decision
to pair Islamic studies with civil engineering may
have been made to appease his family while satis-
fying his own beliefs, but it nonetheless proved
to be fortuitous, maybe even brilliant.

UBL became increasingly militant in his Islam,
and his zeal was buttressed by two events: (1) the
1979 overthrow of the shah of Iran by Ayatollah
Khomeini, and (2) the mujahadin uprising
against the invading Russians in Afghanistan.
UBL fought on the ground with the mujahadin
while simultaneously proving to be their biggest
fund-raiser, bringing the cause as much as $50
million a year, according to the CIA.

The successful defense of Afghanistan em-
boldened bin Laden. He considered himself not
just a religious leader, not just a civil engineer,
but an astute military leader. He returned to
Saudi Arabia after the Afghan war and encountered the third signature moment: Saddam Hussein's August 1990 invasion of Kuwait. He
viewed this development as another opportunity
to further his beliefs and his military acumen: If
UBL's boys could take the Russians, the Iraqis
should be a pushover, right? He offered the
House of Saud an offer he didn't expect it to refuse. He would gather an army of thirty thousand Afghan veterans to fight Saddam, and he
would be the one to lead them.

The Saudis refused, and not only did they refuse but they did the unthinkable: They welcomed an army of three hundred thousand U.S. troops to set up camp on Saudi soil, bringing what UBL considered their degenerate habits of alcohol drinking and sunbathing. So, instead of an Islamic army defending the cradle of Islam, the Saudi government preferred the Americans squatting on sacred soil. Angered, UBL recruited his army anyway, turning his religious-inspired vitriol toward the U.S. infidels and his new enemy—the leaders of his home country. He sent four thousand troops to Afghanistan to begin training, and this budding renegade, a scion of one of the most formidable families in the country, became a significant cause of consternation and angst within the Saudi regime.

The Saudis eventually raided his home, placing bin Laden under house arrest. They didn't seem to know what else to do with him; his family was too influential to make a major stir. The bin Laden family tried to soften UBL's stance, and his refusal caused them to effectively disown him. Then, in late 1990, the Saudi regime got the break it needed when Hassan al-Turabi offered UBL refuge in Sudan. Leaping at the chance to get rid of this thorn in their side, the Saudis encouraged his departure, and UBL eventually succumbed to the pressure and made Khartoum his exiled home.

Many of the Afghan war veterans ended up with him in Sudan, forming the foundation of the al Qaeda army that would enact bin Laden's declared war against the United States. The CIA noted his arrival in Sudan, and we targeted this tall, soft-spoken Arab as a potential threat to our interests. However, his war declaration was not treated with a great deal of seriousness. At the time, we had no way of knowing the man's capacity for evil and destruction. And as it turned out, the congressional hearings following September 11 revealed that the first of bin Laden's operatives reached U.S. shores at some point during 1991, during the time I was watching UBL.

While in Sudan, UBL used his engineering expertise and his money to build roads—*paved* roads, if you can imagine that. His greatest achievement was a seven-hundred-mile road from K-Town to the Port of Sudan. In exchange for this work, the government of Sudan gave bin Laden the monopoly on the sesame-seed export business in the country. Since Sudan was one of the three largest sesame-seed producers in the world, this was a highly lucrative concession for bin Laden.

Bin Laden's tentacles spread throughout the country. He ran a trading company called Laden International. He had a foreign-exchange dealership, a civil-engineering firm. He even owned a company that ran farms that grew peanuts and corn. Another of his civil-engineering works was a new passenger terminal at the K-Town airport. Put simply, the man was living high and mighty in Sudan. Hassan al-Turabi loved him, which means the government loved him. In a financial and recruitment sense, his exile was hugely successful.

It was a perfect relationship. The Sudanese loved UBL's ability to modernize elements of their transportation system, and bin Laden loved the security and freedom he was given in exchange. He used that freedom to conduct his important business—training the al Qaeda terrorists. He was not only the business leader but the religious leader who sat with his workers during prayer as a means of passing along his anger and hate of the Saudi and U.S. government to his less-educated minions.

Not many CIA folks wanted to be in Khartoum at this time; it was dangerous and unpredictable and much friendlier to the bad guys than the good. The only people the agency considered for the duty were males who had a military background, most of them Special Forces retirees like me. PSO officers followed us all over the place. Mostly, they tried to keep their sorry asses from bumping into one another. We'd get tired of seeing these bumpkins, and sometimes we'd play with them a little. They'd follow me to the American embassy and sit in the cars. Inevitably, they'd fall asleep. As I left, I would tap on their car window, startling them to attention. "OK, I'm leaving now," I would say in Arabic. "Better start the car." Then I'd drive them into the desert outside of town, and, boy, did that ever piss them off. We knew the desert better than they did, and they didn't want to follow us out there. We practiced our escape routes, and I drove fast, so they didn't know where I was taking them. They'd follow for a while and figure the hell with it. It wasn't worth their trouble. They'd turn around. I'd finish my route, circling back into town, free of their prying eyes.

I have learned, during fifty years of sneaking and peeking, to sneak, peek, and report with the best of them. My trademark throughout my career, military and CIA, was my reliance on night work. Night was my friend, and I used darkness as if it were a physical entity. I trained myself years ago to operate on a minimal amount of sleep, and I did most of my work between 0100 and 0500 hours, when the rest of the world had long since signed off. I also found that keeping my blue eyes covered with dark glasses, and appearing to be a slight old man with a slow walk and a bent back, got me through some ticklish situations when dealing with the political police of various nations. I kept my alias and cover story in my back pocket for use whenever questioned.

There were times when I worked 24/7 for extended periods of time, breaking occasionally for short (twenty- to thirty-minute) naps. During those precious hours of 0100 to 0500, I found I was able to get right up on the properties owned and occupied by the bad guys—peeking in windows, assessing danger, devising potential operations. I relied on two basic principles: (1) work when everybody else is asleep, or too tired to be alert, and (2) know the block, city, country, or (in this case) desert better than either the police or the subjects of the search. Three words to live by: speed, secrecy, and surprise. To maximize the three, the night can't be beat.

I worked alone at night, and not only alone but as a loner. A perfect example was an operation I conducted in Jordan in 1992 or 1993. The agency needed detailed information regarding the layout of a home in Amman occupied by the station chief of the Iraqi version of the CIA. A staffer stressed the importance of gathering absolutely accurate information regarding this Iraqi diplomat. Following the briefing, I went to work devising a plan to acquire the necessary information.

I decided to conduct the work on foot, using a jogging diversion plan that would take some patience but had the potential to provide huge rewards. Every night at 2 A.M. I jogged a route that took me roughly four miles, from my hotel to the Intercontinental Hotel in Amman. The timing was not random; I knew Arabs were most likely to be asleep at this hour. Although they are well known to stay up to midnight eating and generally bullshitting, by 2 A.M. the revelry always dies down. For the first five or six nights, the Jordanian police tailed me as I made my way through the streets. My route took me past the target Iraqi's residence, but during this time I made no attempt to gather information. I looked straight ahead during the entire run, even when passing the Iraqi's residence, and when I arrived at the Intercontinental Hotel I would drop and do forty push-ups in full view of the Jordanian police parked in front of the hotel. I would not use them until the time was right, but I always carried a night optical device and camera, concealed in my jogging suit.

I am sure now, as I was then, that these police officers were transmitting my route to their superiors. I encouraged their presence by waving to them often. After several nights of this seemingly dull back-and-forth, the police lost interest in this goofy old nocturnal American jogging this insignificant route night after night. They eventually abandoned their pursuit of me, probably out of sheer boredom, and that is when I went to work in earnest. I continued to run the same route every night, though, careful not to act rashly. About a week after they had lost interest, I performed my recon of the Iraqi station chief's quarters.

He had no dog (a big worry), and his car was not bugged. With my trusty Litton night lenses, I took photos of his vehicle and his yard layout. I snapped several photos through the windows of the home. I jimmied the doors and found they were locked. A B and E (break and enter) was not required at this stage of the operation—that would come later. After I concluded my work, this home was the target of an operation I will not describe, except to say his telephone was tapped and listening devices were placed within the home.

My desire to work alone was part of the reason K-Town was such an appealing assignment. Taking risks has always been my forte, and if I break some rules (as I have been known to do), I do not want to implicate any other Americans who might be with me. I'm proud of the fact that

I have never been detected or rolled up on during any of my little forays into the teeth of danger.

My lifestyle in Khartoum wasn't what you'd consider luxurious. Our teams worked together and lived in the same large three-story apartment building in the al-Riyadh section. There were no frills, but each of us independent contractors (ICs) had an individual entrance, which allowed us to come and go without disturbing any of the others. And while bin Laden was driving around in a nice white Mercedes, I was driving an old Russian-made sedan with a big hole in the floorboard. When I drove into the desert, usually at about seventy kilometers per hour, the dust and the heat would boil up underneath the car and come pouring in through that hole.

At one point during my time watching UBL, an agency deputy chief was transferred from a location in South Africa. As part of his indoctrination, the chief of station told me to take him out and give him a tour. We often wore masks to hide our identity—our Americanness, basically—as we drove around Khartoum. These masks were specially fitted to our faces and cost many thousands of dollars to produce. They were intended to make us look black, like the locals, and they looked remarkably lifelike as long as the viewer wasn't up close and personal. My mask created something of a problem, though; its lips were too big, and in order to see through the windshield while driving I had to tilt my head in such a way that the lips were nearly touching the steering wheel. I must have looked like one crazy Dinka as I drove around. The creator of the mask was in Hollywood, though, and I was in K-Town, so what the hell did he care? I also wore the specially designed gloves, and these gloves ran from the fingertips to the shoulder. Remarkable, really, and great cover, but if you're caught wearing one you would certainly be jailed and then booted out of the country as a spy.

So in order for the newly arrived deputy chief to get the full effect of working in K-Town, I set out to teach him the intricacies of wearing one of these specialized masks before I gave him a tour of our escape routes in the desert. Once on the road, the two of us drove along in this old Russian sedan, wearing our masks—him for the first time—and taking in a little Saharan scenery. The hot air and the dust blew into that vehicle like smoke from a bonfire. I noticed my passenger was awfully quiet. Finally he said, in a soft voice I had to strain to hear, "You always drive this fast?"

"Yes, sir."

Pause.

"Where's the floor of your car?"

"Well, some of it's missing," I said. To lighten the mood I added, "Get us some new vehicles and I won't have this problem."

The next thing I hear was an awful sound—*bll-ep*—and the smell told me this deputy chief had thrown up in his mask.

He was angry and embarrassed, but he couldn't

take off the mask for fear of blowing our cover. The sound that came through the vomit and the mask sounded almost like a cry.

"It's your fault," he said. "The way you're driving, you got me all screwed up." When we returned to the station he pulled off his mask. He was covered in his own vomit. The chief looked at him and said, "What in the hell's wrong with you?"

I could see the deputy chief was almost crying now. He flicked vomit off his face and pulled it out of his hair. He told the chief, "He drove across the desert so damned fast, he's got a hole in his car and I got sick. I can't put up with this shit."

The deputy's previous cover was hunting big game in South Africa, a far cry from this assignment. By now it was apparent he wasn't cut out for undercover desert work. The chief just looked at him and said, "You're going to have to get used to this sort of thing." Later that year the deputy was sent back home.

Of course, since Sudan was the one assignment nobody wanted in the early 1990s, I loved it. I wouldn't have traded it for anything in the world. Everything was new; I was one of the very first ICs to work this detail. All these bad guys bumping around, and we were able to come in and scope it out and decide where our safe houses would be and what shape our operations would take. So Khartoum, with its filthy streets and shoddy plumbing and horrendously hot weather, was heaven to me. I've made a career out of being places other people attempted to avoid—Vietnam being the most obvious—and I decided early on in 1991 that Sudan was another godforsaken place I could get to enjoy. There was damned good work to be had in that hell-hole during that period of time. The way I see it, it's pretty simple: If your job is to hunt, surveil, photograph, and report on the movement of bad guys, all in a clandestine manner, then your first order of business is to get yourself where the bad guys are. And for a person of my chosen occupation, no other place in the world offered the opportunities of K-Town in the 1990s.

I approached my agency work as an extension of my Special Forces work. I set out to complete the operations in the same military fashion. I quickly discovered there was a difference in the way ICs who were ex-FBI or ex-DEA or ex-something else approached the work in comparison to ex-SF men. The ex-FBI, for instance, were notoriously bright and often slow. They want to make sure every piece of minutiae is dealt with effectively and conclusively before completing the job. My tactics, as you might imagine, were different. I studied aerial photographs and viewed the targets as I did a military operation. I did not spend a lot of time researching the deep history of the target.

As a result, much of what I know about bin Laden's background I learned long after I operated against him in K-Town. At the time I knew what I needed to know—his daily routine, his attitude toward Americans, his religious fanaticism—and not much beyond that. His childhood or family history was not pertinent to my job. Since then, along with the rest of the world, I have learned a great deal more about this man who became the world's most-wanted criminal.

When I arrived in Khartoum, I was told by the chief of station that bin Laden was one of our targets. "Keep an eye on him," he told me. "We don't know what he's up to, but we know he's a wealthy financier and we think he's harboring some of these outfits called al Qaeda. See what you can find out." I was familiar with bin Laden from agency traffic, but this was the first time I had heard the term al Qaeda. With people like Abu Nidal and the members of HAMAS roaming the countryside, there was no reason to take specific notice of this particular fundamentalist who hated the West. I don't fault the agency for this; there was a long line of bad guys in K-Town alone, and at the time UBL was not at the front of that line.

Bin Laden owned an entire block in the al-Riyadh section of town. He had his three-story residence and various support sites he used for his business enterprises. To the north of his residence, he had a warehouse where he stored his construction equipment. To the south there was a support site that housed the "employees"—men who could just as easily be better described as troops, or believers. It was at this site that UBL led noontime prayer every single day.

Bin Laden, who was thirty-five or thirty-six at

this time, was a loner in Sudan. He drove his white Mercedes alone. Visitors to his home would drive up to his walled compound with their identities concealed by drapes that covered the windows of the vehicle. He had four wives, of course, and numerous children, but the secretive existence he led in K-Town kept us from determining whether they were part of his life there. I have since read that his wives and children were with him in Sudan, but they were never visible to us.

UBL, as fitting a man of such strict religious beliefs, kept a pretty disciplined daily routine. He surely did not miss any prayer session. I did not see him make an early morning prayer session (roughly 0430), but I suspect he took care of that in the prayer area that was inside his compound. He would then head for his companies' support sites and then hit the Arab Bank on Latif Street every workday (Sunday–Thursday) at 0900. He felt so sure of his safety in Sudan that he opened an account in his own name. From there he would visit his support sites again, always hitting his site on the south portion of the al-Riyadh area each day at 1200 to lead his men in prayer. He would then rest at home in the afternoon, and I suspect his resting time included at least one of those wives.

My first task was to find an observation post close enough to his residence to keep an eye on his movements. The CIA didn't know much about him, but it wanted me to be in a position to know more. My people wanted to know

SPECIFIC VIEW OF RIYADH AREA, KHARTOUM, SUDAN, 1992 / 1993.
PHOTOS DAILY FROM OP ⑦ TO UBL SPT SITE (MASTER SKETCH)

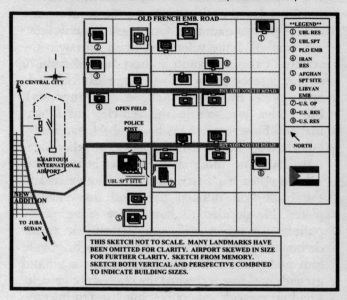

where he went, what he did, and how he did it. They knew about the personal bank account, but they wanted to know more—his associations, his habits, whether he was training some of these al Qaeda folks to do harm to our interests. The chief of station gave me the authorization to spend whatever money necessary to establish an observation post that would enable me to watch and photograph UBL. It was clear this tall, languid man with the curious background was working his way into our radar range.

* * *

The fragile—or nonexistent—social contract on the streets of K-Town would shock even the most-hardened residents of an American inner city. A perfect example: One night I was following one of bin Laden's vehicles through the streets of the al-Riyadh section in a Mitsubishi sedan (agency-provided vehicle, with U.S. embassy plates that were changed with impunity). Bin Laden was not in the vehicle, but it was manned by several of his bearded Afghan bodyguards who were being moved to another location. I was no more than ten feet behind their vehicle when a particularly surly bodyguard in the back seat noticed me—and my embassy plates. He decided to handle the situation by pulling out his AK-47 and pointing it at me through the rear window. I flashed my brights at him and pretended to pull the pin on a hand grenade and toss it toward the rear of the vehicle. I had no grenade, of course, but I stared into the barrel of that AK and gave a little toot of my horn. I waved at the bastard as I turned away from them at the next corner. My attitude was always very simple: Fuck them, my guns are as good as theirs.

Again, jogging was my favorite cover. Going out on foot on fact-finding missions was a habit of mine. I was a pretty good runner, too, despite the junkyard full of scrap metal I carry around in my right foot, ankle, and knee, courtesy of the NVA. There was a health purpose for these trips, of course, but also a practical purpose. At least twice a week I would take that eight-mile run that circled bin Laden's operation, including his house. I needed to get down on the ground and figure out the lay of the land. I needed to know where people were, how they got there, and where they were going. I couldn't do it in a car— a car doesn't allow you the same maneuverability, and it raises even more suspicions if you're creeping along the street, peeking in windows outside of Usama bin Laden's residence. You can't just pull off to the side of the road in K-Town, take out your contraband camera, and start snapping off photographs of somebody's home or business. You have to be smart, and for me being smart has always meant getting out on foot, getting exercise and information.

This is old-school espionage, the kind that has been gradually phased out as technology takes over. At the time I was watching bin Laden, we didn't possess the technology to track a human with a satellite. We can do that sort of thing very well today, but the best we could have done in 1992 was put a beacon on bin Laden's Mercedes and follow his route in that manner. The problem is, a beacon keeps tabs on his vehicle but not necessarily the man himself. To get the job done, you have to get down on the streets and watch him closely enough to discover the pattern of his daily routine.

My jogging routes weren't limited to eyeing bin Laden. We were also interested in the other bad guys over there, and when I ran, I was always on the lookout for vehicles and radio antennas. These nefarious operations had modern communications equipment in this primitive country, and if I could discreetly take a photograph of a vehicle and its antenna, I was in business. The communications guys could look at that antenna and know right away the type of radio and its transmission. Before long, we would be intercepting their communications, which allowed us to know where they were going before they even left. But the way you get to that point is by going out there on foot and doing the legwork, like a cop on the beat.

Bin Laden's ornery bodyguards and their constant vigilance presented a real problem when it came to securing an observation post. I don't scare easily, and I wasn't scared on this assignment, not even while I was jogging and they were training their penetrating eyes on me like lasers. But I also knew enough to know I couldn't get an observation post close enough to UBL's house to set up shop and photograph him on a regular basis. These guards and the logistics of the area made that impossible. History has taught the Afghans to be ruthless and to trust nobody. Bin Laden recruited his followers through prayer, telling them what they wanted to hear. He provided them with an enemy and the justification for their hate. He led them in battle and helped fund their defeat of the Russians, and that bred loyalty. Still, I knew the Afghan bodyguards weren't going to harm me, no matter how much the idea might have appealed to them. The Afghan bodyguards abided by their bosses 100 percent. If the guards knew they would endanger their master's safe haven by bringing smoke on any agency personnel, you could bet your last quarter they wouldn't attempt to kill any American. Sudan was a lawless, on-the-edge kind of place, seriously lacking in proper policing. It was a filthy place, with Dinka tribesmen routinely squatting to defecate next to a Dumpster in the middle of a housing area, but the Sudanese government knew its position with respect to the rest of the world. If an American was injured or killed by one of UBL's people, he would be kicked out of the country immediately for fear of more serious reprisals from the American military.

I knew this. UBL's bodyguards knew this. And most important, they were aware that I was smart enough to understand the situation. This gave me some power in this oddball relationship, but their vigilance made me realize, after several jogging trips, that I would have to find an observation post near one of bin Laden's support sites if I was to accomplish my goal of monitoring his movements and photographing him.

I had to deal with another potential hazard around bin Laden's residence: his guard dogs. He had six or eight desert dogs—big white dogs that were meaner than snakes. These damned things would run up to me as I jogged, just dying to sink their bared teeth into my legs. To combat this, I carried a short iron pipe with me when I ran. A couple of smacks on the snout with that and those dogs decided I wasn't worth the trouble. Whenever I'd run past after that, they'd raise all kinds of hell, but they'd make sure not to get within range of that pipe. I also carried Mace, which would have bought me a few seconds if someone tried anything funny.

Only one time did bin Laden's guards let their curiosity get the best of them. After I passed the house during one of my runs, I heard a car engine start up. One of the guards had decided to follow me in his vehicle, and he stayed about twenty feet behind me the whole rest of the trip. I didn't alter my pace or pay him any mind. I said to myself, "Just suck it up, old boy." You have to have some cojones to do this work. I just kept jogging and he kept driving, as if I were a boxer in training for a big fight. I knew they were packing, but they weren't going to whack me and end up getting their leader thrown out of Sudan, where he was living free and clear. To me, the whole charade seemed pretty boring: following an old man as he jogged up and down a few dirt roads. I was only four or five blocks from my residence, a house owned by U.S. interests, and he followed me the whole way. He saw where I turned in, then sped up and went on his merry way back to UBL's fortress. I didn't care if they knew where I lived. Hey, bring it on. I spent most of my time on the roof of that house, anyway, and during that time I rarely slept more than three hours a day. Chances are I'd see them before they saw me.

Once I reached the conclusion that I couldn't secure a post close enough to bin Laden's residence, my search took me to his support sites. He had a site near the Palestinian embassy, west of his residence near Runway 340 of the Khartoum International Airport, and another south of those sites on South Riyadh Road. I told the chief of station it was not feasible for me to be in a position to watch him twenty-four hours a day, which would have been ideal, but I was pretty sure I could get a spot close enough to the South Riyadh Road site, the site of the noonday prayer sessions, to keep partial tabs on him from there. A little digging and snooping led me to learn that a U.S.-controlled house stood directly across the street from this site. The house was roughly eighty meters to the east from the courtyard where bin Laden led prayer, and the roof afforded unobstructed views of the area. Perfect.

Some U.S. personnel had rented this five-bedroom house for one year but had left with six months or so remaining on the lease. The ambassador gave the OK for me to use the house as an observation post (OP) even though the State Department often fears the CIA. Most State Department people think the CIA's always involved in some shit that's going to get somebody in trouble. I guess we have that reputation.

One of the most important elements of warfare or spying is to seek the high ground. Your goal is always to be looking down on your target or enemy. In Khartoum, this presented a unique problem. Most of the residents deal with the unrelenting Saharan heat by sleeping in the open air, on the rooftops of their houses or apartment buildings. The daytime temperature can reach 120 degrees, and the cooling once the sun goes down is minimal. It can be 100 degrees at three in the morning, and there is no such thing as air conditioning for the locals, except for the very rich. The best most can do is sleep outside, in what passes for fresh air in that stifling heat.

So the problem was obvious: I needed to get to high ground—the top of my building—in order to do my job properly. At the same time, I needed to make sure I was not going to be seen by all these folks sleeping on their rooftops. An American sitting on a rooftop in Khartoum taking pictures of Usama bin Laden wouldn't have left much to the imagination, even at this early stage of his "career." So the first thing I did was build a bamboo cover up there, to allow me to take my photographs undetected. This was a small cover that didn't stand out or raise suspicion. It could have been seen as a rudimentary sun shield used to provide some shade for a local's midafternoon nap. In fact, many of the rooftops in K-Town had them.

Our surveillance had taught us that bin Laden was a creature of habit. He drove many of the same routes every day, from his home to his support sites and often north across the White Nile Bridge and off into a desert town called Omdurman. I would sit at my observation post and wait for him to arrive for noontime prayer. On cue, he'd come barreling down South Riyadh Road in that Mercedes, the airborne dust always his loyal companion, and proceed to the courtyard, where

SPECIFIC AREA OF UBL SUPPORT STATION; A LOCATION VISITED BY UBL DAILY, FOR NOON PRAYER

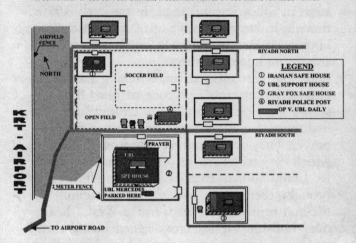

his followers awaited. Every day, right on schedule, he walked directly into the view of my camera.

You have to remember, bin Laden was next to nothing at this time. We didn't have any reason to believe he was capable of mass destruction against the United States. We knew he was angry, we knew he hated us, and we knew he was moving a lot of money around. Those facts were enough for the CIA to find out what he was all about. He went to the bank every day, and you might figure if the outfit knew which bank he used, it would recruit someone within that bank to provide information. Well, by God they did. I get upset when I hear people going on and on about intelligence failures and breakdowns when it comes to bin Laden. We took a lot of the right steps in following this man. In the early 1990s, though, it was impossible to read his intentions. He was an enemy, definitely, but he was not considered an imminent threat.

As I sat across the street, watching and photographing him from the roof of my OP, I observed the way his followers treated him. The first thing they did was gather for their noontime prayer. They sat around reciting from the Koran, which they memorize as one of their obligations to Islam. When that task was complete, he brought them close and started preaching the fire-and-brimstone, death-to-the-West, heavy-duty bullshit. He sat cross-legged facing them, and they sat cross-legged facing him. The followers—twenty to twenty-five of them in this location—sat there mesmerized, silent, mouths half-opened, listening to their messiah as if it were Mohammed himself doing the speaking. It was as if he cast a spell every day at this same time and they—loyal followers all—fell into a dutiful trance.

We didn't have any eavesdropping devices on him at the time, and he was too far away for me to hear his words. He had a very quiet, soft voice, speaking very deliberately and calmly, which pissed me off no end. I always felt if he had been any kind of public speaker I would have been able to hear his words. I used to sit behind my camera lens, about forty meters away, and mutter to myself, "Speak up, you rotten son of a bitch!" I heard nothing beyond a low mur-

mur. Now we're all familiar with that voice, I'm sorry to say. We know how he sounds and how he always seems to have that half-smirk on his face, like he's the only one getting the joke.

My attitude toward bin Laden at the time was based on the way his people responded to him. The more I saw, the more concerned I became. They came up to him like you wouldn't believe. Watching them was enough to make you uncomfortable, even a little disturbed. I knew from experience that the Islamic fundamentalists had the capacity to believe in something to the point where they would give up their lives for it. We don't have that level of belief for anything in our country, and I think part of our shock with suicide bombers and hijackers and people who blow themselves up for a cause is our inability to understand this level of passion and commitment. It seems horribly misguided to us, but it's real and it isn't changing. Our ideas about them have to change to fit reality. We all see what happened: He nurtured this fucking outfit for so long that eventually they started blowing themselves up. They believed—I watched them believe—and the next step after believing is following. To them, following meant doing whatever it took to make his wishes come true, and they ended up following him to the ends of the earth, and beyond.

Bin Laden was forming his al Qaeda cells right then, as I sat there watching and photographing from my little bamboo hut on the rooftop across the street. He had a training camp about twenty-five kilometers north of downtown Khartoum, across the White Nile Bridge in Omdurman. These al Qaeda folks were out there exercising, throwing hand grenades, learning demolition tactics. They were also learning communication skills, like telephone tapping. We knew about it, but we couldn't get out there. The Sudanese didn't allow us to travel north of Khartoum. The police blocked off every exit, prohibiting us from getting a look. The Sudanese were working in conjunction with bin Laden. They were getting their roads and their airport terminal. He, thanks to al-Turabi, was getting the protection he needed to go about his al Qaeda business. Since we didn't have the satellite capability we have now, we couldn't track him. We were out of luck.

We also knew bin Laden built a warehouse twenty-five kilometers east of Khartoum, on the road toward Juba. This little construction project had nothing to do with his so-called legitimate business enterprise. He sent weapons to Somalia during that time. He also sent some of his al Qaeda troops, and they helped train the tribes that were opposed to the U.N. and U.S. intervention in the Somali civil war. It has been established that bin Laden had a hand in the killing of the eighteen U.S. Army personnel on October 3 and 4, 1993, when Operation Restore Hope became a Mogadishu bloodbath.

We'd done our legwork on this guy. I hold firm on this. I photographed him extensively. We knew where he was going; he traveled a familiar route every day. He drove between his support sites, and he drove to the bank, every single day. Bin Laden was a creature of habit, and we had his habits down.

The first step in the operation was to assess just how much of a threat this guy was, based on what information we were able to uncover. The next step was to ask and answer one simple question: What should we do with him?

My job wasn't to make this decision—I was a ground-level foot soldier and information-gatherer—but I could persuade the decision makers with what I knew. Part of that information involved devising a plan to kill the son of a bitch in case someone decided that was the necessary course of action. The plan is called an operations proposal, and it isn't anything special in the outfit. I've written hundreds of them. The ops proposal is simply a way to let your superiors know what's available to them. If it is decided that someone needs to be taken out, the ops proposal puts forth a plan on how to do it.

Remember, bin Laden was not a particularly high priority for us. However, I came to understand and be alarmed by his hatred of the United States. Watching him over time, I reached two conclusions: (1) his followers treated his words with a reverence that bothered me, and (2) it would have been very easy to take him out. In fact, with the way he traveled alone, it would have been a lead-pipe cinch to not only kill the son of a bitch but to get off scot-free afterward.

Someone almost did just that. While I was in K-Town, two Saudis—disgruntled former employees of bin Laden—tried to kill him. Believing UBL was in an Omdurman mosque, they broke in with AK-47s and sprayed, killing about fifteen people. Undeterred by the incompetent Sudanese police, these two would-be assassins sped across the White Nile Bridge and all the way to bin Laden's residence on the east side of K-Town. Total distance, twenty-five kilometers, and they were never even slowed by the police. In fact, they killed two Sudanese officers while driving past a police post about a half a block from our house. Roughly twenty minutes later, we heard a flurry of gunfire from the direction of bin Laden's residence. We later discovered that bin Laden's big Afghan bodyguards waxed those two Saudis. Bin Laden, wherever he was, escaped unscathed.

As that story indicates, bin Laden's house was very secure. But when he traveled, he traveled alone. He was a cowboy during his stay in Sudan, flying around in that big-ass Mercedes like a guy without a worry in the world. He was a fucking king. As I watched him every day and got to know his routine, I grew more and more convinced that his freewheeling ways made killing him an easy task.

I offered a proposal to kill bin Laden during one of his regular trips to a support site outside of town. The proposal called for the use of just one diversionary tactic. One of our vehicles would follow UBL's car while another approached from the opposite direction. The ap-

proaching vehicle would veer into the oncoming
lane and slam into bin Laden's car, stopping him
in his tracks. That was the diversion. Once bin
Laden was stopped, the driver of the trailing car
would get out and shoot him right then and there
with a silenced MP-5, an automatic weapon that
makes absolutely no noise. It would have been
that easy.

This wasn't an elaborate plan—two vehicles,
two men, one weapon. We didn't need any advanced tech support—especially given where we
were. Sudan was, and is, a rowdy, lawless place,
with little in the way of rules and not too many
people interested in following them. The PSO,
the bumbling group of Sudanese security in
charge of watching us, was so inept that killing
bin Laden would have been met with almost no
obstacles. To get away clean, after the fact,
would have been almost too easy.

Unfortunately, at that time permissions to
kill—officially called Lethal Findings—were
taboo in the outfit after a 1976 executive order
signed by President Gerald Ford. That ban has
been lifted in light of September 11, but in the
early 1990s we were forced to adhere to the
sanctimonious legal counsel and the do-gooders
at the agency. If we could have foreseen bin
Laden's intentions, if we could have guessed at
the depth of his crazed hatred of America, we
could have given this bastard the send-off he deserved. For the cost of one 10-cent bullet, all of
that tragedy could have been averted. One 10-

cent bullet and he would have been dropped in K-Town's dusty streets, a place that befits this man, and left to rot like the dog that he is.

I could have killed him myself. Every day I photographed this man, and every day it occurred to me that I could just as easily shoot him as point my camera at him. I was close enough and had a clear enough shot to take him out. We could have saved ourselves the horrible consequences of his terror. It might not be right to say it, but had I thoroughly understood his capacity for destruction, I would have done it on my own. I would have shot him, and then my bosses would have said, "Goddamnit, Billy, he got shot while you were watching him. What the hell happened?"

"Don't know, sir," I would have said. "The man got shot out of the blue. That kind of thing happens in these parts."

And what was the cost of our bleeding-heart "humanitarianism"? Thousands of people would still be alive without it. And two U.S. embassies in East Africa, one Kobar Tower, one USS *Cole*, and the World Trade Center would be operating full swing to this day.

How I wish we—and I—had known this man's bottomless capacity for evil. I wish we had known that one day the dust from those streets in Khartoum would become dust from the World Trade Center.

Dust to dust.

Damn.

CHAPTER 9

In December of 1993, I sat at my CIA reporting site in northern Virginia listening to a briefing that outlined the particulars of my forthcoming assignment. The CIA briefing officer ran down the various details I would need to undertake the upcoming trip to my favorite haunt: Khartoum.

My newest assignment, as it turned out, was a lot like my last assignment. Once again, I was headed for K-Town, the notorious center of the bad-guy universe.

K-Town was rapidly becoming my home away from home, and I listened to the briefing knowing this assignment wouldn't require much area study on my part. Scarcely two months earlier, in October 1993, I returned to the United States from one of my many assignments to Sudan. I had spent more than a year there during the period of 1991–1993, all in six- to eight-week segments.

I didn't like what I saw around me in Sudan—

a highly deprived country with people dying on the streets—but I wasn't there to make political or moral judgments. I was there to work, and K-Town remained fertile ground. In a twisted way, I had grown fond of that city and the possibilities that existed on its dusty streets and inside its ramshackle buildings. I knew the city and the surrounding desert like I knew my own name. I knew its landscape and idiosyncrasies better than the people who lived there, and better than the people who were paid to keep me from doing my job.

The silly hide-and-seek game we played with the mostly incompetent PSO had not changed since my time watching UBL. They still fumbled along, attempting to keep track of us. We continued to befuddle them with our ability to elude them in their own backyard. It was business as usual.

As I sat in Virginia and scanned the cable traffic, I took note of the names of the IC team members who were on the ground in K-Town at this time. I would soon be leading this four-man team, and I was heartened to see familiar names on the list. I knew them all and had worked with each of them previously. This was a good team, a team I looked forward to joining. Don, an ex-police officer from the Midwest, was a particular favorite of mine. He was absolutely superior in searching out persons and vehicles of interest.

All of these men were well versed in the peculiarities of this unique assignment. Put it this way: I was confident each of us working the

K-Town detail could lose the PSO tails within twenty minutes of heading for the desert.

We could not rest during our time in Sudan; our whereabouts were of constant interest to the government and, by extension, the PSO. The PSO was headed by President al-Bashir, and the force knew exactly which Americans working out of the U.S. embassy in K-Town worked strictly for the State Department and which— like me and the other ICs—worked for the CIA under State Department cover.

In fact, a list of the CIA workers had been published, in English, by the government of Sudan. My cover name could be found on this list, but it was of little concern to me. They could know about me, and they could know about my true job, but that knowledge was offset by my proficiency in identifying tails or spotters who were on the government payroll. I was rarely surprised during my numerous stays in K-Town.

However, I did spend one day at an out-of-headquarters site studying and reading the recent traffic from the station at Khartoum. I educated myself as to the current operations and got a handle on what I would face once I arrived. I scanned the usual number of international terrorists who had been given sanctuary by the Islamic government led by al-Bashir and strongman Vice President Hassan al-Turabi. Again, no surprises.

Following the day of study, I arranged my travel and briefed my IC bosses in northern Virginia. I departed Dulles International Airport at

1700 hours on December 13, 1993, and slept
like a newborn most of the way across the At-
lantic Ocean. In Frankfurt I boarded a direct
flight to Khartoum. Just before touchdown, as
the big Lufthansa bird passed over the White
Nile River, I looked over the city and saw the familiar brown haze hanging over Khartoum and
the surrounding desert. This haze seemed permanent—almost like airborne silt as it hung over
this city located at the rushing confluence of the
White Nile and Blue Nile. The arid climate and
frequent windstorms *(haboobs)* created this condition and earned K-Town the nickname "Brown
Town" from our people.

Upon deplaning in K-Town, I proceeded
through customs, using my black diplomatic
passport in the way it was intended: to avoid
search of my bags and roll gear. Several very tall
and black female customs agents—there were
very few males in this position—towered above
my five-foot, nine-inch frame. They glared close-
mouthed at me, clearly unimpressed with my
status as a "foreign diplomat." The male Sudanese customs officers sat in their offices, drinking the local coffee, while their ladies conducted
the baggage searches.

The customs personnel slapped a big white
chalk check mark on my baggage, and I continued with my head high and eyes straight,
through the diplomat exit of the customs section.
I had been through this before, and my familiar-
ity with the routine kept me composed and un-
concerned. There was a reason that each of the

CIA ICs chosen to work in Khartoum were male, and most were former Special Forces men: For an American, being in Sudan was like being behind enemy lines.

I joined my two awaiting IC mates and we proceeded to the parking area at Khartoum International Airport. We got into our 1980 Toyota Land Cruiser and drove toward our diplomatic villa in the al-Riyadh section of town.

K-Town had a curfew of 2330, and by the time I cleared Sudanese customs, gathered my bags, and found my friends, it was 2400—midnight. To nobody's surprise, we were stopped at the corner of Airport and Riyadh roads by a few *jundis*—Sudanese soldiers with no rank. They informed us we were out after curfew, which wasn't exactly news to us.

By this time, my Arabic was good enough to conduct an elementary conversation with these men, so I asked them how they were.

"OK," one replied, then he quickly said, "Give me cigarette."

We always kept extra packs of cigarettes in our vehicles for this very purpose. We had a pack for each of the soldiers manning the curfew stop point, and everyone was happy. They smiled at us and waved us through.

The rest of the short drive was uneventful, with the view outside the Land Cruiser's windows in keeping with the monochromatic theme of Brown Town: one brown building after another. I'd say 95 percent of the buildings in K-Town were as brown as the air that hung above

them, because 95 percent were constructed of
Nile River mud, supported by a minimum of
concrete.

As we approached our villa, I noticed a very
tall Sudanese guard, as black as a human could
possibly be. He smiled upon seeing me, and I
recognized him immediately as a man I had
worked with previously. Hamid was about six
feet seven inches tall and weighed in at 180
pounds, tops. I remembered him as being a
Christian Dinka tribesman who was vehemently
anti-Arab. And, therefore, vehemently against
the Islamic government that led Sudan.

I called to him from the passenger side,
"Keefak, Hamid."

(How are you?)

He replied, *"Quaise jiddan, schokaran Siad
Beeley."*

(Very good, thank you, Mr. Billy.)

I mention Hamid because he was someone I
learned to trust, and in this business trust was a
valuable commodity. There was no guarantee
that a man guarding our villa was not aligned
with our enemies. In fact, Hamid told me he was
routinely grilled by the PSO, who asked him to
report on our intentions and actions. He refused
to yield to their interrogations, and Hamid and I
developed a fine relationship. He was yet another
memorable character from my many days in foreign lands.

I had a room on the lower deck of our villa, a
spacious building that was more than adequate
for our purposes. The ICs and I sat together until

about 0200 that morning, the morning of December 14, discussing the situation in Khartoum. I asked about all the staff members assigned to K-Town, as I was friends with all of them. I passed out some candy and other goodies that I had brought with me; nothing remotely resembling goodies was available in K-Town.

At first light the next morning, roughly 0530, I awoke from a couple hours of sleep and took my morning jog. The weather, even in December, was excellent. I made sure to take my iron dog-pipe with me as I left the villa and headed to the north. These K-Town dogs barked and yelped all night, every night, and generally were a big pain in the ass. They seemed smart, though; I'll give them that. It didn't take long for the message of my iron pipe to circulate through the desert-dog commo network, because while I jogged they remained off and away from me, satisfied to yelp from afar.

At 0700, back in the villa and cleaned of the city's dust by a quick cold shower, I proceeded to drive with my fellow workers to the U.S. embassy. I noticed a white Toyota truck—the preferred vehicle of our friendly neighborhood PSO—pull out behind our Land Cruiser. We turned to the north on Airport Road and headed to the embassy complex on Latif Street. The white Toyota followed us and remained one hundred meters or so behind us as we crossed under the railroad bridge into Khartoum City.

We turned off into the alley across the street from the embassy and parked. I noticed the PSO vehicle, with its three uniformed personnel. I waved at them, letting them know we were in on the game. These men simply glared at us, attempting to look menacing despite their lowly stature as PSO *jundis*.

One of the passengers was inside the cab, while the second was standing in the bed of the pickup, his hands on the roof of the cab to maintain balance. He stood there with his AK-47 slung over his right shoulder and burned me with his stare.

Business as usual in K-Town. I was comfortable here, and ready for the work to come. To this point, I thought to myself as I walked toward the front gate of the embassy, this was just another tour of duty in Brown Town.

I would soon find out how wrong I was.

I entered the embassy's front gate and displayed the Khartoum Embassy badge my teammates had secured from the regional service office from one of my previous tours. The three outer guards were also Dinka types: very tall with million-kilowatt smiles. All were cordial, and each also knew me as *Siad Beeley*. The Marine guard, another man who knew me from my previous tour and greeted me as "Mr. Billy," buzzed me in after I displayed my badge.

There was a reason I was "Mr. Billy." Only first names were used as a means of protecting the use of last names. Cover names—our real first names attached to a cover surname—were easy to mix up, even for veteran officers.

The elevator took us to the third floor, where the Regional Service Office and CIA offices were located. My fellow ICs knew the combination to the entry lock, and we entered. This was like old home week for me; the office manager was a bright female who had worked for the agency for more than twenty years. She was one of the most efficient people I had ever met, just phenomenal. She often ran the whole office whenever the chief of station (COS), Cofer Black, was absent. There was no deputy chief at the time, so the chief was a very busy man.

Black was a hulking, no-nonsense type who had been in the Directorate of Operations (DO) for more than twenty years. He was a stand-up guy and one of the best at running an overseas station.

As soon as we made our way into the CIA office, Cofer called a "stand-up"—a meeting of all staff and IC officers. We filed into his office and he introduced me to the group of agency officers. Of course, the frequency of my visits to K-Town meant I knew almost all of them from past work.

After the formalities and introductions were completed, the boss informed our IC group of our mission: In this city of one million souls, we would be responsible for finding and fixing none other than Ilich Ramirez Sanchez, the man known far and wide as Carlos the Jackal, the world's most famous terrorist.

While in the Washington, D.C., area, I had been prepared for the possibility that the forty-

four-year-old Carlos would be a top target of
this particular mission. I sat there, letting Black's
words sink in, and the adrenaline began to
course through my veins. Like combat adrenaline, it set my mind on high alert.

Oh, man, I thought, *I love this shit.*

My work photographing and watching bin
Laden was fascinating, but its importance would
only be made apparent to me as time went on.
UBL wasn't much of a big fish at the time, but
this was different. This hunt could make a career, and the stakes meant it had to be conducted
with the utmost stealth.

This was Carlos.

This was the biggest fish.

We were about to chase a legend.

I listened closely as the latest information was
passed along by the outgoing IC team leader,
who discussed how his team had broken down
the search of the city for Carlos. The outgoing
team leader repeatedly referred to his map and
some imagery as all of us—including Cofer
Black—sat in on the briefing.

Carlos's code name was "Charlie" for radio
transmission purposes. He also had a cable code
name, which I will not mention.

Carlos the Jackal had one hell of a history. I
was familiar with his file and his legend, and I intended to learn as much as I could as fast as I
could.

Despite Carlos's high profile, this was not an
easy mission. The man had not been pho-

tographed for more than ten years, which meant that any available photos would be of dubious worth in our quest to identify this scoundrel.

The rise of Carlos the Jackal in the early 1970s coincided with my retirement from the U.S. Special Forces. At that time, I paid cursory attention to this man whose catchy name always seemed to be preceded by the term "major terrorist." The term "terrorist" was new in those days, and there was little reason to believe it was going to become part of the world lexicon in a matter of a few short years. There was also little reason for me to believe Carlos would ever become a major figure in my life.

As Carlos became more well known and his exploits gained worldwide notoriety, I was no different from any other news-driven American in my growing curiosity regarding this scoundrel. I tracked the rise of his criminal profile and the mythology that surrounded his antics. Being a Special Operations man, I was interested in the audacity and brazenness of his operations, not to mention the uncanny manner in which he always managed to avoid capture.

My interest is not to be confused with respect, but I was well aware that Carlos had been involved in many spectacular—and spectacularly successful—terror events. Until the early 1990s, of course, thoughts of Carlos in a professional sense did not occupy much of my time. I had other badasses on my screen and was busy around the world, dealing with them.

* * *

A Venezuelan-born Marxist, Carlos allied himself with the Palestinian cause at a young age. He joined the violent Popular Front for the Liberation of Palestine and committed his first outright act of terror on December 30, 1973, when he entered the home of prominent London businessman Edward Seiff, a leading fund-raiser for Jewish charities and the head of Britain's Zionist Federation, shot him through the upper lip with a revolver, and escaped. It was a bungled job; Seiff lived, but the legend of Carlos was born.

I read as much as I could find, including David Yallop's book *The Hunt for the Jackal*. I learned Carlos was bold and audacious. He was also extremely lucky.

Carlos is believed by many to have played a role in the kidnapping and murder of Israeli athletes in the 1972 Summer Olympics in Munich, but that connection has never been confirmed. However, it is undisputed that his reign of terror began in the early 1970s and thrust him into international prominence. His crowning achievement took place in December 1975 in Vienna. Carlos, along with a group of Palestinian and German terrorists, broke into the offices of an Organization of Petroleum Exporting Countries (OPEC) meeting and kidnapped eleven Middle East oil ministers. The terrorists killed three during the attack, then commandeered an airplane and flew to Algiers. The terrorists released the hostages in return for $20 million ransom. Carlos and the kidnappers surrendered in Algiers but were released by a sympathetic government within a few days.

The OPEC operation was suicidally bold and irrationally successful; Carlos reportedly kept $10 million and turned over the other $10 million to the Popular Front for the Liberation of Palestine. A year later, he again made international headlines with his involvement in a Palestinian hijacking of a French airliner headed to Entebbe, Uganda. That operation ended with the famous Israeli commando raid.

In 1982, a terrorist team cell led by Carlos attempted to blow up a nuclear reactor in France. Their rocket-powered explosives were not able to penetrate the facility's concrete walls, so the attempt failed. Later that year, the French arrested two of his accomplices, Magdalena Kopp and Bruno Breguet. Carlos wrote a friendly letter to French authorities, specifying the consequences if the pair was not released.

When the French refused, Carlos went to work. He bombed a French facility in Beirut. Two weeks after that, he bombed a French passenger train from Paris to Toulouse, killing five. A couple of weeks after that, his men assassinated a French embassy worker and his pregnant wife in Lebanon, then bombed the French Embassy in Austria, then a restaurant in Paris.

The man knew how to play the retribution game. Eventually, after Carlos's numerous acts resulted in twelve deaths and 125 injuries, the French government released Kopp and Breguet. Kopp joined Carlos at his headquarters in Damascus, Syria, and the two were married in 1985.

Carlos strung together twenty uninterrupted

years of shadowy activities, outright terrorism, and unbelievable good luck. He referred to himself as a "soldier, and I live under a tent," when in reality he was smoking the most expensive cigars and drinking the most expensive liquor. He acquired a worldwide reputation almost as fast as he ran through girlfriends. Carlos was twenty-six years old at the time of the OPEC operation. Everyone in the world who read a newspaper knew his name.

The CIA had been trying to get Carlos for more than fifteen years. He laid low in Lebanon, in the Bekaa Valley, then went to Syria and eventually Yemen. Our intelligence indicated he spent time in Tunisia. He lived in lawless countries that ignored—or, in some cases, encouraged—international terrorists. He also wore out his welcome in many of these free-for-all countries, which seems indicative of his character.

Before finding sanctuary in Sudan, Carlos lived in Jordan, and before Jordan it was Syria. The noose had been tightening around the neck of Carlos. Syria wasn't in the habit of forcing anyone out of its country, no matter how unsavory, but they ultimately ran Carlos's ass out of there. The United States was pressuring them, picking up telephone calls and telling the Syrian government, "Look, we know he's there. If you don't do something about it, we're coming to get him." They got tired of that, so they told Carlos to get lost.

While in Jordan in 1993, Carlos married a young Jordanian Arab and made her his second

wife. Carlos was a self-anointed Islamist, which
allowed him to take four wives and numerous
concubines. His stay in Jordan was short; he
wore out his welcome there pretty damned
quick, and he was asked to leave in August 1993.

He was living in Khartoum at the pleasure of
President al-Bashir, but he was officially welcomed and overseen by Vice President al-Turabi,
who took personal responsibility for Carlos the
Jackal, in much the same way that he took pride
in sheltering Usama bin Laden. I learned all this
as I researched the available material on our new
favorite target.

I had to fight my creeping suspicion that Carlos the man wasn't nearly as imposing as Carlos
the legend. He carried himself as a badass, dress-
ing and acting the part. But even taking into account the probability of exaggeration, the
available evidence on the Jackal indicated he was
a shallow mercenary whose love for money ran
at odds with his Marxist beliefs. It was my contention that ideology took a backseat to his desire to get drunk, chase pussy, and show the girls
how tough he was.

I had done my homework on this miscreant, so
I did not walk blindly into K-Town in December
of 1993, when Cofer Black told me we were embarking on a mission to cease the terror activity of
Carlos the Jackal, once and for all. I knew a good
deal about the abilities and habits of this man.

I was prepared to help the CIA take him
down.

My study led me to believe his lifestyle would

catch up to him. His womanizing and drunken debauchery had caused him to fall out of favor with many Arab nations. His actions caused them, rightfully, to question his professed belief in Islam and devotion to the Palestinian cause. I was convinced his growing lack of discipline would cause him to make a misstep that would allow us to nab him, once and for all. I wasn't sure when it would happen, but I knew if the agency gave us the resources to make it happen, chances are Carlos would go down a lot sooner than later.

His elusiveness was due in large part to the sanctuary he was provided by the governments of various Arab and Communist nations. He moved from one to another, from Yemen to Lebanon to East Germany, hidden under the cloak of policy. Believe me, when a government decides to allow the entry or exit of a particular individual—especially a wanted one—they have the means of hiding that person. The problem, as I saw it, was simple: The Sudanese allowed Carlos the Jackal too much freedom of movement. Other governments did not allow him to move around so freely, and most of them harbored him on the condition that he stop committing terrorist acts.

After the meetings on my first full day in K-Town, Cofer Black pulled me aside and said, "Billy, this is the man. You've got to get this guy." At that moment, given the gravity evident in his voice, I knew the agency was making this a top priority. We would get the resources, and then we would get the Jackal. I wanted to be the guy who caught this asshole.

Carlos lived on rumor, and he had a whole network of people whose job it was to spread rumors of his whereabouts. This was a big part of the Myth of Carlos—he's here, he's there, he's everywhere. When enough people hear enough different scenarios, nobody knows what to believe. He understood this phenomenon and cultivated it, working it to his advantage.

His only followers were those on his payroll. He developed a sort of Robin Hood persona, the little guy who took from the money-hungry capitalists to aid the downtrodden. The myth that surrounded this guy was unbelievable. It got to the point where whatever happened in the world, well, it must have been Carlos. You could have a bank robbery somewhere in the world and someone was bound to say, "Must have been Carlos." He loved that notoriety. He lived for it. This was what he was all about.

It was evident the money he received from the OPEC kidnapping had softened him some. In addition, his safety depended on his continued retirement. There was talk in the first Gulf War that Saddam Hussein had recruited Carlos to come out of retirement and help his cause, but that didn't advance beyond rumor. By the time I arrived on the case, it seemed the feared Carlos was pretty much out of the killing business and into the fun business.

We set out to end the party.

* * *

I spent my first two full days in K-Town discussing the plan of action with the departing IC team leader. Cofer Black's directions were exceedingly simple: Get out there on the street and find Carlos the Jackal. It was an easy order to follow, but a difficult one to complete.

During the final days of 1993, and the early days of 1994, the other ICs and I searched Khartoum, as secretly as possible, for the Jackal. Remember, the CIA was not in possession of a photograph of Carlos that was less than ten years old, so identifying this man in a city of more than one million people was like searching for a shadow in the dark.

For more than two weeks, we searched secretly and fruitlessly for Carlos. It was obvious we needed some sort of break to get ourselves closer to the Jackal, and during the second week of January, our break came from an unexpected source: Carlos himself.

We received information that the Jackal, from an undisclosed location in Khartoum, had placed an overseas phone call to one of his trusted bodyguards, a man I will call Tarek.

The message Carlos passed to Tarek concerned some trouble Carlos had gotten into with the Khartoum police. Apparently the famous Jackal had waved his omnipresent pistol in the face of a Khartoum shopkeeper during a drinking binge, and the local police had thrown his ass in jail. He informed Tarek on the phone call that his release from jail had been arranged by Hassan al-Turabi, the country's henchman vice president. Carlos requested Tarek come to Khartoum ASAP to assist him and act as his personal bodyguard.

Tarek agreed. In the meantime, we received information from an overseas CIA station that Tarek was a well-known thug and bodyguard for the upper-echelon members of the Palestinian terror groups in the earlier days of Carlos.

This station, which will remain unidentified, cabled all known information regarding the relationship between Carlos and Tarek. They went through their files and came up with some additional intelligence that could serve our purposes: information regarding the Jackal's new Jordanian bride, Lana Abdel Salam Jarrar. Included in that cable was a photograph of the woman I shall call Lana ASJ.

This woman was approximately twenty-three years old and beautiful. Through the cable traffic, we surmised that Lana ASJ had accompanied her man to Sudan. I studied the photograph carefully, as did all members of the CIA station in Khartoum and every IC. We knew there was a high probability she would become a factor in our search.

The information regarding Lana ASJ was solid and helpful, but there was one piece of news from the station that trumped all: This obliging station had in its possession a recent photograph of our man Tarek.

To us, this was a gold mine. We had knowledge that Tarek was on his way to meet Carlos from another Middle Eastern country, and now we had access to a photograph of the bodyguard to accompany this knowledge. In the intelligence business, this was a major development. The photo of Tarek was sent immediately by secure fax to the Khartoum station.

When the photo arrived, everyone in the station studied it carefully. Tarek appeared to be Caucasian, approximately forty years old with wavy white hair. He was extremely fit, with the look of an accomplished bodybuilder. We were notified that Tarek carried an Iraqi diplomatic passport, among others, but his appearance did not fit the description of an Arab.

When Carlos moved to Sudan, he was granted not only sanctuary but immunity. He was protected, as were his associates. This proved to be a problem for us when our agents were unable to track Tarek when he arrived at Khartoum International Airport. He was treated as a distinguished guest of Vice President al-Turabi and did not disembark the plane in the same fashion as a normal passenger.

We guessed that Tarek descended the aircraft's stairway and was whisked into a waiting car. We knew he was on the plane. We had people waiting for him. We just couldn't get to him, which meant that Tarek left the airport without a tail.

This was not a major setback; we just had to work a little harder to get to Tarek. On January 20, 1994, I was with an IC named Greg as we searched the normal places a Caucasian would visit in Khartoum. This, we knew from experience, would not be the bazaar area, or *souk*. We

went into the Meridien, a hotel in which I had
lived for several weeks in 1991, but found no
Tarek and decided to take advantage of the opportunity to eat a good meal in the hotel's
restaurant. Although Greg and I weren't having
any luck with our hunting, we knew one fact beyond a shadow of a doubt: The seven-foot Dinka
who cooked at the Meridien made a fine plate of
lazana.

Our bellies full, Greg and I left the restaurant
and entered the lobby area, near the reception
desk. It was there that both of us dropped our
eyes on Tarek sitting at a table, working a cross-
word. This man was an exceptionally large human being, with muscles bursting out of his
T-shirt. Damn, this guy's muscles had muscles. I
couldn't imagine how many hours a day a person
would have to pump iron to look like this guy.
Well, Carlos wanted protection, and it looked
like this character could provide it.

Since both of us were savvy and experienced
operators, we walked calmly past Tarek as if he
were just another Sudanese sitting in the hotel
lobby. I knew Greg had seen him. Greg knew I
had seen him. But we couldn't draw attention to
ourselves or Tarek for fear of blowing our covers, and we couldn't ask any of the hotel clerks
whether Tarek was a resident of the hotel. That
would not be a wise move, because both of us
knew the hotel clerks in Khartoum—especially
hotels like the Meridien—worked for the PSO.

This brings up an important point: One of the
keys to being a successful spy is keeping your

head down and minding your own business. If you keep your head down and bend your back a little bit like you're old and not able to harm anybody, you can acquire a lot of information. I'm old anyway, so it comes naturally to me, but even in my younger days I found I could get a lot of work done by minding my own business. That's exactly what Greg and I did instinctively as we identified Tarek and headed out the door of the Meridien.

Simple Surveillance 101: If you spot an adversary, don't let your eyes linger on him or her. Never allow your eyes to meet the adversary's eyes; this is a dead giveaway. Always have something to do; in a hotel lobby, for instance, have a paper to read, and when you spot the target, immerse yourself in that paper like it's the last thing you'll ever read. Never allow your eyes to flit around; this is a sure sign of amateurish surveillance. Always work in teams and have a signal rigged to indicate to your partner or other team members that you are "on target." This signal could be simple, as simple as the act of sitting down. You can point your shoe toward the target and your partner can see the line of your shoe and follow it to the target. The partner stays completely away from the target and takes a position that will enable him to watch the target depart the area. What you need, generally, is a vehicle with a license number, which gives you all you need to reacquire the target.

Greg and I left the lobby and walked purposefully toward my CIA-issued Toyota Land

Cruiser to await Tarek's exit from the Meridien.
Before long, we watched Tarek exit the hotel and
walk to a white 1990 Toyota Cressida. We noticed the license was a Khartoum civilian plate,
number 1049.

Tarek drove that Toyota Cressida toward the
Blue Nile River and immediately made a U-turn
before we had even started our tail. Tarek had
been involved in espionage for years and was
well versed in deterring surveillance. This is why
Greg and I were wary of attracting attention, so
we chose to put a loose tail on Tarek's vehicle. It
was about 9 P.M., and the haphazard traffic system in Khartoum did us no favors. In short order, Tarek was able to drop us and have the road
to himself.

Our disappointment in losing Tarek was nothing compared with our excitement in finding him
and identifying his vehicle. We reported our findings to Black, the Chief of Station. Our work on
January 14 had been fruitful: We had a fix on an
automobile that was being driven by Carlos's
handpicked bodyguard.

The search was narrowing and gaining focus.
Now all we had to do was find that white Cressida on the streets of K-Town.

The hunt for Tarek was over.

The hunt for the Jackal was just beginning.

CHAPTER 10

We searched sixteen hours a day, every day, for the 1990 Toyota Cressida driven by Tarek. The four of us organized our search by separating the city into two sections. One IC worked the Khartoum city side while another remained on the Riyadh side, searching diligently for the white Cressida.

Don, the ex–police officer, was among the best at conducting searches. His particular specialty was license plates and vehicles. He was able to read plates and identify vehicles from a great distance, and his phenomenal memory made him a superstar in our field.

Obviously, communication was vital. The station had a commo system with a VHF repeater, which allowed each of us to be armed with a portable radio. It was imperative that we know one another's whereabouts at all times. The transmissions were encrypted, which prohibited others from intercepting our words. Those with-

out the encryption key would hear a rushing
noise and no words, like the sound of wind
howling through a cave.

We knew Carlos had a reputation for being a
playboy, so we concentrated our hunts in areas
that would appeal to that side of him: the Meridien and Hilton hotels, the Greek and German
clubs, and the Diplomatic Club. These clubs
served booze, even though alcohol was strictly
forbidden in Sudan by Islamic law. The Diplomatic Club, located along the banks of the Blue
Nile River, threw a big shindig on the first
Thursday of each month, holding a disco-style
dance on the eve of the Friday Muslim holy day.
The bash at the Dip Club was known to be a
gathering place for Iraqis, Egyptians, and some
Libyan diplomats who enjoyed dancing and hard
liquor.

This was an obvious place for us to search for
Carlos and his buddy Tarek, so on the first
Thursday of February, Don and another IC
named Greg staked out the Dip Club parking lot
in search of the 1990 Toyota Cressida known to
be driven by Tarek. While there, Don contacted
us on the radio to say they had spotted the Cressida.

Don and Greg followed the car as it departed
the Dip Club and headed north toward K-Town,
about fifteen kilometers away. The traffic along
this highway south of the city—particularly on
Thursday nights—was routinely jammed with
any vehicle you could possibly imagine. The
weekly celebration filled the highway with carts,

vehicles, donkeys—you name it. But the most striking obstacles were the tall black Dinkas who walked the roadway, dressed to the nines in dark clothing. There are no lights along the roadway, and as you might imagine, any nighttime driving along this corridor must be done with extreme caution. That's a roundabout way of explaining why Don and Greg lost sight of the Cressida shortly after it left the Dip Club parking lot, and also why they were unable to find it later that night.

Immediately upon receiving the encrypted communication regarding the spotting of the vehicle, I set up a vehicle-surveillance post at the intersection of Airport Road and the turnoff that led to the al-Riyadh section of Khartoum. I was awaiting the arrival of the Cressida in order to take up the tail as soon as the vehicle passed my clandestine position.

The vehicle, however, did not pass. This led me to make two educated guesses based on my knowledge of the area: (1) the Cressida bypassed the major thoroughfares by cutting directly through the desert, and (2) Carlos and his companions lived in the New Addition area of K-Town, the most logical place for a person of his VIP standing. Taking shortcuts through the desert, where no road existed, was a common practice; I did it many times.

The New Addition consisted of sixty-five numbered blocks. Only odd numbers were given to the blocks—that is, the street that followed Thirty-fifth Street was Thirty-seventh Street then

Thirty-ninth Street and so on. Every home or apartment in K-Town was surrounded by a mud wall that stood three meters tall, in the traditional Arab style. What did this mean to us? It meant that on the night of February 3, 1994, there was a high probability that the Cressida was parked within the walls of one of the many homes or apartment buildings in the New Addition, with the front gates closed. Carlos was a wanted man, and this trusted bodyguard was summoned from afar for reasons that went beyond his rippled physique. This guy was expected to keep his man safe and sheltered from prying eyes like ours.

We could not drive repeated circles around the New Addition that night; each of the homes and apartments are more than one story, and many residents combat the heat by sleeping in the open air on the roofs of these buildings. To drive around and around would be to invite observation and suspicion. As a result, we quickly abandoned our search for the Cressida.

In contrast to my days in combat, I was forced to develop patience in this line of work. Progress was measured by different standards, more incrementally, and by those standards we had moved forward in our search for the Jackal. We made two major discoveries—the bodyguard and his vehicle.

We were getting closer and closer, inching toward giving Carlos the unceremonious ending he so richly deserved.

* * *

BLOWUP OF THE NEW ADDITION; INT AIRPORT; RIYADH AREA OF THE SOUTHERN SECTIONS OF KHARTOUM, SUDAN

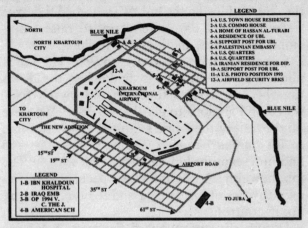

Espionage is not always glamorous. It has its moments of suspense, danger, and adrenaline highs, but most of the time it is a calculating profession, with success predicated on stealth, cunning, and perseverance. Often the adrenaline rushes are followed by days and days of watching and waiting. Those days—the down days—were as important to the mission as the moments of excitement. I was forced to remind myself of those truisms often during the six days following the sighting of the Cressida in the Dip Club parking lot.

Between February 3 and February 8, we concentrated our search for the Cressida in and around the New Addition area, especially south of the crossroad of Airport Road and the main route to the south of Khartoum. This was the route we believed Tarek followed after leaving the club and losing the tail. We searched this area sixteen hours a day, careful not to raise suspicions. We looked and looked—nothing.

Sometimes the tedium gets to you. The job is filled with fits and starts, and when the down time became wasted time I found myself frustrated and anxious to get the job done. We were on the trail of the world's most famous criminal, so it's human nature to want to move fast—to find him, capture him, and be done with it. The job, however, moves at its own pace, creating its own rhythm of excitement and tedium. We were not in control.

During the afternoon of February 8, at approximately 1400 hours, I was explaining some of the finer points of film developing to Greg in the sixth-floor photo lab in the U.S. embassy. We had just mixed chemicals for black-and-white photography development. Like everything else, developing film in Khartoum had its own peculiarities. The water in K-Town came out of the pipes at roughly ninety degrees Fahrenheit, so we had to put ice cubes in the chemicals to cool the mix to its desired temperature of sixty-eight to seventy degrees.

I explained this to Greg and was interrupted when I heard my call sign—"Batman"—coming through via encrypted call from Don.

"Batman, I have spotted the white Cressida parked adjacent to the Ibn Khaldoun Hospital on Nineteenth Street in the New Addition."

A buzz ran through me. The hair on my arms

stood at attention. I needed to think clearly and quickly. The fruitlessness of the past five days was pulverized by the excitement I sensed in my friend's voice.

"Roger," I said.

I took a breath and continued, "Take up a surveillance position. Photograph the passengers if the vehicle departs, then tail the vehicle to its destination."

I turned to Greg and said, "He might have him. Let's go."

We hurried to the embassy elevator and waited impatiently as it descended the six floors to the ground. Once outside, I glanced at a parked PSO Toyota pickup.

Its occupants were both asleep. If I had had time, I would have laughed out loud. It was perfect.

We climbed into my Land Cruiser and drove south, careful not to speed or otherwise drive in a manner that would invite attention. We checked again to make sure we had no tail and then turned toward the hospital.

One of my constant companions was a fully loaded Canon F-80 35 mm camera with a Canon 300 mm lens. This equipment was in my vehicle, in a small lockbox I had welded into the floor of the backseat. I asked Greg to get the camera out of the box and attach the lens.

We reached the Ibn Khaldoun Hospital in roughly five minutes. Don was parked facing the west side of the hospital, one block from the front entrance. As I turned from Airport Road

onto Nineteenth Street, I spotted the empty
Cressida about one block to the east of the hospital, parked the wrong way, its front end facing
toward me and Airport Road. I nosed into position on the side of the road, on the same block as
the Cressida, approximately twenty meters away.
I parked in a way that ensured a clear and sure
shot of the Cressida with my Canon 35 mm.

Greg and I were in five-by-five radio commo
with Don. This terminology means it was
crystal-clear—"I hear you five by five" is the best
there is.

With our positioning perfect—us on one side
of the Cressida, Don on the other, the hospital
entrance between us—we waited.

Less than ten minutes had passed since Greg
and I had been idly discussing photo chemicals in
the embassy lab. That's the way it worked—all
the dull hours of searching followed by an immediate rush to action. We couldn't plan for this,
but we had to be ready for it. We were measured
by our ability to think on the fly; any hesitation
could result in a missed opportunity. We didn't
know what we would discover. Tarek could be by
himself. Tarek could be with Carlos. Carlos,
somehow, could be alone. We were also on the
alert for Lana ASJ, the Jackal's Jordanian bride.

As I surveyed the situation on Nineteenth
Street in the New Addition of Khartoum on that
afternoon in February 1994, a plan formed immediately in my head.

Across the street from our position, on the
same side as the Cressida, was a cigarette stand

with one Dinka vendor sitting on the road's dirt shoulder.

I told Greg, "You see that guy there selling cigarettes? Here's what we're going to do. Walk over to his stand and check it out. Purchase some cigarettes and then hang around until someone approaches the Cressida. When someone comes, start raising holy hell with the Dinka in a way that makes sure you attract the attention of every single person on this street."

This was to be our diversionary plan, and my friend understood its purpose immediately. "Roger," Greg said with a definitive nod.

Greg then walked across the street and began viewing the wares at this Dinka's small stand. I got out of the Land Cruiser and walked to the front of the vehicle. It was typically hot—ninety-three degrees. I flung open the hood and peered at the engine as if I had mechanical problems. Two Americans could not simply park on the side of a K-Town road and expect to go undiscovered for any length of time. The appearance of engine problems might buy us a little time and provide a little cover.

I left the hood open and climbed into the front passenger seat on the right side of the Land Cruiser. I placed the 300 mm lens through the driver's-side headrest. This enabled the headrest to act as a fixed platform for the lens and camera. I then adjusted the lens so that it was zeroed in just perfectly on the exit door to Ibn Khaldoun Hospital.

As I was adjusting the lens, a lady I immediately recognized as Lana ASJ—the Jackal's pretty young wife—appeared in the lens. If I hadn't been so intent on doing my job, I would have considered it magic the way she materialized before my eyes.

She was alone, and I quickly snapped off four or five photos of her. She walked toward the Cressida, and as the shutter was clicking I heard Greg begin the task of raising hell with the Dinka salesman.

"Goddamn you, you cheated me, you son of a bitch!" Greg yelled. "Who the fuck you think you're messing with? I'm going to beat your ass!"

I glanced up at this commotion, thinking what a good job Greg was doing. The Dinka arose from his seated position on the ground and just kept going and going. Damn, this guy was about seven feet tall. Greg kept up his tirade, though, and when I looked back toward the hospital entrance I saw a large Caucasian man exiting with a young black man. My first thought upon looking at this big Caucasian was *Goddamn, that's Carlos the Jackal*. He was fifteen years older and probably forty-five pounds heavier than in his previous photos, but my heart raced and, once again, the hair on my arms stood on end. This had to be him.

I kept him in the viewfinder and clicked off frame after frame as the two men walked toward the Cressida. Greg continued his verbal assault on the Dinka, causing the man I believed was Carlos and his black companion to stop and stare. *Perfect*. A few more frames clicked off with this big man occupying dead center in my viewfinder.

Greg interrupted his verbal assault on the Dinka long enough to look across to me and say, in an insistent stage whisper, "Billy, do you see the man?"

I said quietly in return, "I have him."

And I did. This roll of thirty-six-exposure film was rapidly clicking to a conclusion, and if my experience told me anything, it was that I was working on the photography equivalent of a perfect game. A simple diversion, an empty roll of film, and the most famous terrorist in the world—could it get any better than that?

I took note of this heavy man's appearance: roughly six feet tall, weighing somewhere between 210 and 220 pounds, very neatly groomed, hair turning reddish. He had a brown bag slung over his shoulder. For certain, this was not a camera bag but was most likely a weapons' bag. I also spotted a leg holster near his right ankle. He was wearing a shooter's jacket—a sleeveless vest with several pockets in the front.

This man, suspected but not known for certain to be the Jackal, stopped as he neared the passenger door to the Cressida and watched again as Greg finished his diatribe against the Dinka. By this time, the utterly confused Dinka vendor was getting a little pissed off. He had no idea why my friend was berating him, and he'd had enough.

Lana ASJ got into the driver's seat and closed the door. The man believed to be Carlos got into

the front passenger seat. The black escort placed
a large envelope—one that looked as if it might
contain X-rays—into the backseat behind Carlos
and returned to the hospital. (We later learned
Carlos suffered from venereal warts, and some
reports indicated he was undergoing tests to determine the cause of a low sperm count.) Since
they were preparing to leave, still completely un-
aware of our presence, I quickly and covertly
transmitted to Don, still parked a block away.

"Follow them to see where they go," I said.

"Roger" came the immediate reply as the
Cressida pulled away, toward Airport Road,
where it turned right (south) with Lana ASJ driving. Tarek was nowhere to be found, only the big
man in the passenger seat and the pretty young
driver.

Upon seeing the Cressida depart, Greg
reached into his pocket and pulled out Sudanese
pounds in the equivalent of twenty U.S. dollars.
He handed the money to the dumbfounded
Dinka, took a quick look to judge the man's reaction, and then jogged across the street to join
me in the Land Cruiser. The poor Dinka looked
at the money and shook his head. As the amount
he held dawned on him, the dumbfounded look
turned to joy. *What the hell,* he must have
thought. If he had to take a little abuse, so be it
for that kind of money.

I dropped the hood of the Land Cruiser,
started it up, and headed west along Nineteenth
Street. I pretended not to notice as Don drove
past to set up his tail on the Cressida.

I was absolutely beside myself with glee and anticipation; I was almost certain the big Caucasian was Carlos the Jackal. I said a few prayers to myself that the film was good, that the photos were good, that the camera had done its job.

I whooped to Greg, "Fuck, my friend, I believe we just got Carlos the Jackal on candid camera."

Greg was as elated as I was. We exchanged congratulations on what appeared to be a job well done. We had set up a hasty but effective diversionary, which Greg had run to absolute perfection. It took *cojones* to unleash the assault on the unsuspecting—and undeserving—Dinka, but Greg had pulled it off and kept it going long enough and well enough to attract the attention of our main man Carlos.

I attempted to call Cofer Black, who had been at a meeting with Ambassador Donald Petterson when we departed the embassy at 1400. I looked at my watch—1550. All this took place in less than two hours. I received no reply from the chief, but his office manager told me Black had returned to his home. I tried again, this time attempting to contact him by VHF portable radio, but again there was no answer. Using a telephone was out of the question in Khartoum—an instrument of the Sudanese government, it was surely tapped, and very scratchy also.

Unable to reach our boss, we drove immediately to the U.S. embassy. We exited the Land Cruiser and walked past the PSO Toyota pickup, where the two PSO *jundis* sat there, in the same position as when we left. This time, however, they had advanced from fully asleep to half-asleep. A smile on my face, I saluted them from a distance.

My camera, with the treasured film residing safely therein, was inside a gym bag I always kept slung over my shoulder. Greg and I stopped briefly at the station office. I told the office manager we would be in the photo lab on the sixth floor for at least an hour. I revealed no further information.

Once in the lab, I carefully checked the chemical mixes to ensure temperatures were proper. My heart raced. This was the most important roll of film in my career. *Don't screw this up, Billy boy.* I added ice to the mixes and saw the temperature decrease to its desired number. It bounced between sixty-eight and seventy degrees—perfect.

I extracted the film in the total darkness of the lab. I took no chances and cut no corners in undertaking this task. I carefully followed the timing rules of development and generally treated this film like a fragile newborn. To the second, I withdrew the film from the lightproof containers at the proper time.

Over the years I had acquired a real affinity for film development. I worked hard at it and got to be pretty damned good at it, to be perfectly honest. As you develop photos, you know when film goes through the first solution and comes out of the second solution, all it takes is exposure to the red darkroom lightbulb for you to know whether you have good images. If you have handled the film properly, the figures or targets will jump right out of the solution and into your eyes. This moment—the moment the strips of film are pulled out of the solution by tongs and held up to the light—is the moment of truth in film developing.

As I pulled the negatives out of the solution and lifted them up to the light, I held my breath in anticipation of what I might see. All alone in that room, facing such an important moment for both me and my country—maybe even the whole goddamned world—I waited to see whether I had been able to capture an image that would lead to the arrest of the world's first celebrity terrorist.

When I removed that film from the second bath and held it up to the red light, a buzz raced through my body and the hair on the back of my neck stood up. "Damn, oh damn," I said out loud to myself in that room. There he was, Carlos the Jackal in reverse, jumping out of those negatives as if he were standing next to me.

This film was more than simply positive identification. This film contained thirty great shots of a man many believed to be the most dangerous and elusive man on earth.

The Jackal had avoided the camera's lens for more than a decade, and here I was, standing in the darkroom with a silly grin crossing my face, confident the Jackal had performed his last criminal act of terror.

This was it, the end of Carlos, staring back at me from the film I held in my hands.

I was out of my mind with elation. The trepidation and anxiety that took hold of my body
while I developed the film vanished, replaced by
the giddiness of a job done pretty damned well. I
cleaned the film, then cut it up for enlargement.
No fingerprints would be on this film.

Don called with an update of his surveillance
of the Cressida. He followed the group from the
hospital until it pulled into a gated apartment
complex on Thirty-fifth Street, abutting Airport
Road.

The man suspected to be Carlos got out.

His lovely bride got out.

The couple went inside the front gate of the
complex and walked up one flight of stairs, into
an apartment.

We now had a potential residence to accompany our pending identification. It was shaping
up to be a historic day's work.

I continued to call the COS from the film lab,
but I couldn't reach him. I called Greg into the
film lab to view the results of our actions. One of
our Khartoum case officers—a full-time CIA
staffer—had been monitoring transmissions. He
asked me, via portable radio and in encryption
mode, "Hey, Batman, what do you have?"

"I might have something great," I replied.

Those words prompted the case officer to
come to the photo lab. With his surveillance of
the Cressida complete, Don arrived at the photo
lab at around the same time. He had just learned
to develop film and was aghast at the speed I was

using on these precious photos. He was convinced I was not babying the film properly, and I figured he was just glad he wasn't responsible for these photos.

Once I washed and dried the eight by tens—all thirty-five of them; one was not as clear as the others—I was amazed at the clarity and quality of each photo. The photos of the beautiful Lana ASJ were ideal. From thirty meters, her features were perfect with the 300 mm lens, as good as photographs can get. The twenty-five photos of the black escort and Carlos—at least the man suspected to be Carlos—were similarly perfect.

I continued in my attempts to reach the Chief of Station. At approximately 1630—roughly two hours after taking the photos—I placed another unanswered call. Where was he? Convinced we had no other choice, Greg and I headed to Cofer Black's home in Khartoum. At this point, the communicating needed to be done in person. Black needed this information now. The *world* needed it now.

His home was a twenty-five-minute drive from the embassy, toward the Blue Nile River, past the home of Hassan al-Turabi, past the Chinese embassy, into the eastern side of the al-Riyadh area of K-Town.

The guard at the COS's quarters told me Cofer was inside, so we rang the bell. This after-hours call was unusual for a couple of ICs, and after a few rings Cofer Black came to the door in a pair of running shorts, a T-shirt, and running shoes.

We had gotten him off his Nordic Track machine, interrupting his method of blowing off steam after a long day in Khartoum.

Cofer Black looked at me and said, "What in the hell is up, Billy?" It was obvious he knew nothing about the activities of the past four hours.

"Well, sir, I have some photos I want you to see."

He washed the sweat off his hands and looked at us quizzically.

I handed him a manila envelope. He looked at the first of the thirty-six photos, a close-up of Lana ASJ, and said, "Jesus Christ, Billy, this is Carlos's goddamn wife. Where in the hell did you get these?"

"Look at the others, please, sir."

Black then withdrew several photos we believed to be Carlos the Jackal. He stood there, dumbfounded, his jaw hanging open for a few seconds.

He said nothing. When he had looked at all the photos, he laid them on his dining-room table.

Finally, he asked quietly, "Who took these?"

Greg answered.

"Billy took them, Cofer, about two hours ago at the Ibn Khaldoun Hospital."

"Holy shit," Cofer said, the disbelief gradually receding from his tone. "I believe this is Carlos the Jackal, but we must be certain."

Black thought for a moment and then turned to me.

"Billy, find one of the ICs we can send back to the States and have him leave *tonight* with these photos."

I gave him the name of an IC whose tour was up on this coming Friday, three days away. Black nodded and said, "Take care of it."

We were directed by Black to keep three of the suspected Carlos photos and one of Lana ASJ, then send the rest by "safe hands" in a diplomatic pouch.

"Have this IC who takes these identify the agents who meet him at Dulles," Black said, "for he will turn over the photos at that time to the agency duty officer for positive identification."

We scrambled. We packaged the photos properly and received the required letter for movement of a diplomatic pouch. All of this was done by a State Department individual. We transported our departing IC with this package to the Khartoum International Airport for an 0200 departure on February 9 on a Lufthansa flight direct to Frankfurt. From there he would travel to the Washington, D.C., area.

It all somehow came together and our photos departed. We held our breath, hoping against hope that this had been our friend Carlos.

On the morning following our euphoric and (we hoped) successful operation against Carlos and his lovely bride, we gathered in the chief of station's office at 1000 hours to discuss the next steps. Everyone from the station was present, and I briefed the chief, all staffers, and the other

ICs on exactly how the photo session had gone down.

We debriefed our boss Cofer Black as to all times, decisions, and actions that occurred during the finding of Carlos and the subsequent setup at the Ibn Khaldoun Hospital. Black had to brief Ambassador Petterson ASAP, and he wanted all of us to be on the same sheet of music before he met with the ambassador.

We rehashed every phase of the completed action, in much the same manner we did following combat action. Don was commended for his sharpness in noticing the Cressida as it was parked on Nineteenth Street. Greg and I were lauded for the diversionary and the great photos that resulted from our hasty but effective plan.

I was aware that Cofer Black sometimes thought I took action without bothering to consider approval or repercussions, but he did not mention this. I knew he would have rather I contacted him before taking the photos of someone we believed to be the world's number one terrorist. Since time was of the essence, and we did not know whether the opportunity would present itself again, he did not chastise me for completing the action.

In preparation for the next round against the Jackal, we discussed what we knew about this man and his wife. It wasn't much, but we thought we knew the following:

• Where they resided in the New Addition area of K-Town.

- The vehicle they used for transportation.

- The identity and physical appearance of Tarek the bodyguard; we suspected he lived in the same apartment building as Carlos and his lady.

- That neither Carlos nor his friends were aware of our impromptu photo session.

What we did not know made up a much longer list:

- Carlos's intentions.

- Carlos's communications system.

- His standing with the government of Sudan.

- His future terror actions, if any; we hadn't the foggiest idea whether he had a network of associates masterminding his future terror plans, or whether he had retired from the killing business.

- His inner circle in Sudan, presuming one existed.

At the end of the briefing, Cofer Black summoned his sternest voice and warned us that he did not want surveillance on the Jackal, at all, until he ordered it. If we saw the Cressida or any of the individuals involved (Lana ASJ, Tarek, or Carlos himself), we should report this to him immediately as a spot report, which he would relay to the headquarters people in the U.S. They, of course, would become intensely and immediately interested in this operation as soon as a positive identification was garnered from my photographs.

Of course, I had my own opinions on how we should decide the fate of the Jackal. To no one's surprise, I felt we should place surveillance on this fat-ass jerk. In fact, I requested that we save the government a lot of money and kill Carlos. This could have been done easily, and it could have been completed in such a manner that we could have pinned it on some other government that had a bigger ax to grind with the has-been playboy.

Alone with Black in his office, I put forth this proposal. He quashed it outright. He told me, in no uncertain terms, that he would make the decision on future action, and only after determining if the person was, indeed, Carlos the Jackal. I had to get my opinion out there, though, if only to cement my reputation as someone who spoke his mind.

Cofer was now required to visit the ambassador, and he told me to remain in his office while he did so. After passing along the information to Ambassador Petterson, Black returned to his office and summoned me and my IC partner to the ambassador's office on the second floor of the U.S. embassy. The ambassador gave us the usual platitudes, congratulating us on a job well done.

Meanwhile, we waited anxiously for the important information: word from CIA headquarters on whether the individual in the photos was, in fact, the Jackal. We tracked the IC who carried the photos and found he landed at Dulles at 1400 hours on February 9. He was then taken to CIA headquarters, where the heads of two divisions viewed the photos.

In an unexpected twist, the IC was then told to take the photos in a diplomatic pouch to Jordan—ASAP—where the Jordanians could positively identify the subject.

The IC—accumulating frequent-flier miles at a stupendous clip—traveled back to Frankfurt, then on to Amman, Jordan, with the photos in a dip pouch. The Jordanian intel identified the man in the photos as absolutely—definitively and without question—Ilich Ramirez Sanchez, Carlos the Jackal. The IC was told to return to the United States with the photos, for the expressed purpose of turning them over to the chief of the Near East (NE) division.

Back in K-Town, we were jubilant. There was handshaking all around. We were well on our way toward doing what so many countries had been unable to do: capture Carlos the Jackal.

Undaunted by Black's dismissal of my plan of action, I asked for and received aerial photos of the entire city of Khartoum. I proceeded to devise a plan of action should minds change and we be detailed to take Carlos out of this world by force. Of course, with the Clinton administration in office at the time, I held out little hope that such action would be authorized.

We continued with our other duties around

the city. Plaudits and congratulations for finding
and fixing the Jackal rolled into the chief of station from around the world. Carlos, the target of
our affections, continued to roam the streets of
K-Town, going through life half-drunk, clueless
of our recent accomplishments.

On February 19, 1994, I was ordered back to
the United States, where I was congratulated by a
few of the top-deck people at the agency. I traveled back and forth from Washington, D.C., to
Africa four times in the next two months.

All remained quiet on the Carlos front, but it
was simply a matter of time before the world's
most-celebrated terrorist would meet his demise.
We had fixed his identity and his residence. More
than fifteen years of searching had been con-
densed into one unprecedented afternoon of
work.

The hunt for the Jackal was over.

The takedown was about to begin.

CHAPTER 11

In the days following our photo session with Carlos, I was faced with long airport layovers and interminable flights to and from various worldwide airports. This provided me with plenty of time to contemplate how I had come to be involved in this remarkable action against the world's most-wanted man.

Khartoum was not my first assignment linked to Carlos. During the summer of 1992, the agency directed me to Bregenz, Austria, to support an operation in that city involving a reported lieutenant of the Jackal. I shall not name this individual, or mention his code name, but will simply refer to him as Target. This Target billed himself as a trusted confidant of the Jackal.

In my opinion, Target's story is an illuminating example of the legend of Carlos and how it grew to outlandish and completely unwarranted proportions.

In preparation for the trip to Bregenz, I was briefed for several days, then directed to fly to Zurich, Switzerland, and drive to Bregenz. The Swiss countryside, I must say, is magnificent in the summertime.

Bregenz is located just across the Swiss border, on the far eastern tip of Lake Constance. One tunnel out of the city leads to Germany, another to Switzerland, still another to Liechtenstein. The four-country crossroads is a storybook location, like a scene from *The Sound of Music*. When it comes to working conditions, it beat holy hell out of Khartoum. There weren't nearly as many badasses roaming the landscape, however, which made it unlikely that I would be given repeated assignments to hunt cretins along the shores of Lake Constance.

When I arrived, I met with the other ICs. We received further instructions and completed the job. I can't reveal any details about the operation, but I can provide some colorful sidelights that offer a glimpse into the mythology that wrapped itself around Carlos and refused to let go.

After the job was complete, I sat alone eating a Wiener schnitzel at a bar/restaurant in Bregenz when Target happened to walk right in the door. He was alone, although we knew he had a girlfriend in Bregenz. He was dressed to the nines, and immediately I identified him as a dapper fop.

I ate and minded my own business, but I watched and listened as he started a conversation in German with a couple of young ladies at the bar. Watching his smooth and unctuous approach, I thought to myself, "This guy appears to be nothing but a bullshitter, but he's quite an accomplished ladies' man." When I completed my great meal, I departed without saying a word to anyone, and I never looked directly into the eyes of this self-described "trusted lieutenant" of Carlos the Jackal.

This man, like much of the mythology surrounding Carlos, was fraudulent. This man was on the agency's payroll, and he was feeding us phony information on Carlos's whereabouts. He claimed to be the Jackal's top man, his right hand, and he was as full of shit as a man could be. We didn't know his information was bogus when we were tracking him in Bregenz, but as time passed we got smart to his game. Part of our enlightenment came as a result of seemingly innocuous observations like mine at the restaurant. Sometimes it's the simplest chance encounters that provide the most revealing glimpses into a person's true character.

In 1994, after I had photographed Carlos in Khartoum and we had successfully nailed down his place of residence, our man "Target" told his handler to let our people know Carlos was in Lebanon, preparing to do evil things. That was how Carlos operated, with bold lies and paid liars. Inside, he must have known his strategy of constant misdirection couldn't work forever.

There was a figurative cease-fire on the Carlos front following our photo session in front of Ibn

Khaldoun Hospital. The demand for immediate action against Carlos was not forthcoming. The chief of station was worried that too much activity near or around Carlos's apartment would surely give away the fact we had this man in a box.

As a result, the CIA instituted a two-month "stay away" policy regarding Carlos. I was traveling back and forth from Washington, D.C., to Africa, and the ICs who remained in Khartoum kept occasional tabs on the Jackal. I surely did not agree with this policy—I would have opted to set up twenty-four-hour surveillance—but I tempered my anger with the educated assumption that Cofer Black and his people had a plan in their back pockets.

Even though we were confident Carlos's terror career was on life support, there were still indications the legend remained hale and hearty. On February 21, 1994, thirteen days after the photo session, I was alerted to travel to Cairo, Egypt. Information had been reported that Carlos would be relocating to Cyprus and passing through Cairo International Airport along the way.

As the team leader of a four-person team, I assumed a position at the Africa terminal of the Cairo airport. I watched the exits of several flights from Sudan but saw nobody resembling our famous friend. This detail lasted for six mind-numbing days. Total take: no one. Another red herring on the hunt for the Jackal.

A month later I was detailed to return to Khartoum. Once there, I suggested to Cofer Black that I search for a suitable stationary OP from which our team could observe Carlos the Jackal's apartment and track his movements. He agreed, feeling it was time to accelerate the hunt and get this bastard off the street once and for all.

The search took time. I looked through the month of March and into April. Had Carlos, his wife, and bodyguard decided to leave K-Town secretly, or if the government of Sudan had decided it was in its best interests to spirit him out of the country secretly, we would have been hard-pressed to discover this information beforehand.

This heel-dragging carried into mid-May. Carlos was still in K-Town, as reported by ICs and others who saw him out drinking and making the playboy scene. He attended one of the Dip Club discos during the month of May, and we knew as much. Occasionally an IC or other station member would see Carlos riding around in the Cressida. All sightings were reported via VHF portable radio, but the surveillance got no more active.

The search for a proper observation post dragged as well. Then, on May 20, the outgoing IC team leader and I visited an old apartment building on Thirty-seventh Street in the New Addition. The north side of this building afforded an unimpeded view of Carlos's apartment, which was one block or roughly 120 meters to the north, on the corner of Thirty-fifth Street and Airport Road. This old apartment complex had several occupied apartments on the lower floors, but we received information that the top floor, six floors above Thirty-seventh Street, was dilapidated and unoccupied.

The outgoing team leader—the man whose place I would take—was extremely aggressive in his search for an OP. He and I were eager to bring Carlos back to the front burner; it was evident the United States would not take action against him anytime soon. We figured a good OP, overlooking Carlos's place of residence, would bring us information on his comings and goings, as well as the identities of those who paid him visits. Our photographs combined with this information, we assumed, would return Carlos to the front burner PDQ.

One of the occupants living in a lower-floor apartment in this run-down building served as its manager. My partner and I spoke to him and expressed our desire to take a look at the top-floor apartments that faced to the north. When he looked at us sideways and asked about our intentions, we had a cover story ready.

"We are making a survey for the U.S. government," I said, "and we need a place to store our equipment."

I immediately sized up this character as a greedy type, and it was apparent he could smell U.S. dollars about to come his way. This little twerp didn't care if we came from Mars and were planning to use the building to conduct alien experiments; he was sniffing cash and jumped at the chance to rent out more apartments of this sad old building.

He showed us the top floor, and it was really trashed. Pieces of ceiling had fallen out, there were holes in the walls, and the floors were filthy. There was no toilet or running water. Horrendous as it might be, though, we saw one thing: a superb view of the entire front portion of the building where we knew Carlos and crew resided.

As I stood inside that shabby sixth-floor apartment, I looked across toward the subject building and actually saw the white Cressida parked behind a closed gate. I feasted my eyes on the clear view of the front gate of the apartment—the only entrance for visitors entering the building, and the only exit for those departing. I knew that very second this had to be our observation post.

The first order of business was to barter to rent the apartment. The greedy bastard of a manager asked for $700 a month for this wreck. The manager, of course, saw this as his chance to sock it to the *Ameerikee,* so he capped off his demands by asking for three months rent, in advance.

We weren't about to lose this place, so we told him we would most likely take the apartment. We then hurried to report all of this to the chief of station. When I described the location and the view, making sure to emphasize my sighting of the Cressida, Cofer Black nodded his head in complete agreement. This was the place.

The next step was to come up with $2,100 cash right away. The three of us—the outgoing team leader, Cofer Black, and I—forked over $700 apiece so we could lock up this deal *now*. The rental would be approved by the agency, but that approval would take forty-eight hours. We didn't have forty-eight hours, and losing this prime piece of real estate (for our purposes, anyway) was not an acceptable option.

We returned to the apartment and handed over the $2,100 in U.S. dollars to this little creep, whose eyes sparkled as he pocketed the cash. We wasted no time—another IC named Dennis and I moved into the dump right then and there. To say this place was a dump was an understatement. As bad as it was in terms of living conditions (the worst), it was just as good as an observation post. On the day we turned over the cash, this place became our home.

It would remain our home for the next three months.

We moved into the apartment on May 28, 1994, and it didn't take long for Dennis and me to realize we had made absolutely the correct choice. On the first afternoon of our "occupation" I viewed Carlos himself, as he departed his apart-

OVERVIEW OF KHARTOUM, SUDAN METRO

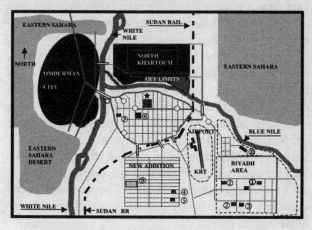

```
LEGEND
① HOME OF UBL                    ⑦ IRANIAN EMBASSY
② SPT SITES OF UBL ▬             ⑧ U.S. EMBASSY
③ PHOTO SITE UBL                 ⑨ KHARTOUM CEMETERY ▭
④ APR OF CARLOS                  ⑩ RESIDENTS OF U.S. PER. ▬
⑤ VS. CARLOS OBSERVATION POST       NAMED CONDO APTS.
⑥ NATIONAL CAPITAL OF SUDAN      NOT TO SCALE
```

ment building, with Tarek at the wheel of the Cressida.

The living conditions were beyond reprehensible. There was, of course, no elevator in the building. Its odd construction forced us to be low-key during the move-in process. The open staircase ran along the outside wall of the building, which meant we had to get our surveillance and communications equipment up these stairs without the world reporting what we carried to our old friends with the Sudanese PSO. Just the fact that a couple of white men were moving into an apartment this disgusting would have been of interest to the PSO.

The lack of privacy caused Cofer Black to assign only two of us to work this OP detail. No others were to visit, for fear of giving away the entire operation. Despite Sudan's camera ban, we were about to stock this observation post with the world's best gear for a photo-and-watch operation against Carlos the Jackal. We had to lug this gear up the six flights of stairs, and we had to do it secretly. The outdoor stairwell left us no choice but to conduct the move-in at night.

On May 26, two days before we moved in, a video lens arrived by means of dip pouch from the United States to the Khartoum station. This thing was the biggest lens I'd ever seen. It weighed 140 pounds and was approximately twenty-four inches in diameter. Dennis and I were taught how to operate this powerhouse video-camera lens, which had a reaching power of 4,000 mm, 3,200 mm, and 1,700 mm. The 4,000 mm lens could reach out for more than one kilometer and get identification-quality video of license plates, people, or whatever you might need to have identified.

An all-camera 1,000 mm Nikon lens and an 800 mm all-camera lens were both delivered in the same shipment. This was great, spare-no-expense equipment. There was a drawback, however: The gear weighed in at a total of 250 pounds. Add the heavy tripod and we were looking at gear in excess of 300 pounds. We had all this gear plus three tripods to hump up five flights of stairs, clandestinely, on the night of May 28.

Dennis and I began the move-in after dark, when the night activity of the New Addition area had quieted. We carried the 4,000 mm lens practically step by step, one of us on each end of the case, careful to remain quiet and calm. It was hard work, made harder by the need to regulate our breathing to make sure we attracted no attention.

Eventually, step by step and piece by piece, we carried every last bit of equipment into the observation post before first light. The two of us cleaned the garbage and crap—I'm talking literally here—off the floor of the apartment. We set up two folding army cots, placed black curtains over the bare windows and—*voilà!*—the next morning, May 29, 1994, our Carlos the Jackal OP was open for business.

Between this ragged building and Carlos's building, coincidentally, was a residence that housed the higher-ranking officers in the K-Town police force, as well as their families. It was also located on Thirty-fifth Street, across from our OP but far enough to the west to keep from obstructing our clear, line-of-sight view. This police residence was about the same height as our OP building, causing us some concern that an OP in that building could actually be directed toward us.

In this line of work, you have to take into account every possibility.

Of course, as in any surveillance, I prepared a Daily Journal/Log, recording all times and events. We had the following items to assist our surveillance during the first days:

- Two portable Motorola radios with encrypted channels
- One Canon F-80 35 mm camera with a 300 mm Canon lens
- One Canon OES 35 mm camera
- Tripod for camera and lens
- Binoculars and a shooting spotter scope, for a range of a thousand meters
- Sketch pads and sketching materials
- One 2-kilowatt portable generator
- Two GI-type folding cots with covers
- Food for two days
- A Bunsen burner, to cook rice and heat water

During the first days of our surveillance, I initiated the daily journal, which was kept in the U.S. embassy. The journal was set up as follows:

EXAMPLE OF LOG OF DAILY JOURNAL

DTG*		TIME	EVENT OCCURRING	PHOTO	INFO
1.	06/01/94	0600	Black Civilian Enters Compound	No	No
2.	"	0900	Carlos into Vehicle (Sits Only)	Yes (4)	All
3.	"	0910	Carlos into Building	No	All
4.	"	1100	Lana ASJ / Carlos into Veh	Yes (6)	All
5.	"	1101	Carlos Opens Gate/Lana Drives	Yes (2)	No
6.	"	1103	Lana Drives Both Depart to the West on 35th St	Yes (2)	All
7.	"	1330	Vehicle Returns Lana Drives Carlos Opens Gate/Veh Park	Yes (3)	All
8.	"	1400	Other Resident on Balcony	Yes (2)	No

*DTG = Date/Time Group

Since photography would provide proof positive of Carlos's presence in Khartoum, I prepared a photo log, which gave us the means of recording and identifying each photo and the time taken. It was set up as shown:

DTG*	TIME	ROLL #	EVENT / OF WHOM	FRAME #
1.	06/01/94	0900	3RD	Carlos Sits in Cressida	1–4
2.	06/01/94	1100	3RD	Carlos/Lana in Veh	5–10
3.	06/01/94	1102	3RD	Carlos Opens Gate	11–12
4.	06/01/94	1103	3RD	Carlos/Lana Depart Lana Drives	13–14
5.	06/01/94	1330	3RD	Carlos/Lana Return Lana Drives—Inside Gate	15–17
6.	06/01/94	1400	3RD	Unidentified Resident on Balcony (non-C's Apt)	18–19

*DTG = Date/Time Group

To me, this careful recording of the photos not only appeared very professional but would prove invaluable in identifying Carlos and his associates, as well as those who came into contact with them.

Keeping logs was sometimes frowned on by the people for whom I worked at this time. They weren't especially interested in their men carrying notes on their person. Dennis and I devised a method to eliminate this possibility. We kept notes in a notebook, then transferred the information to a log inside the U.S. embassy. It meant, of course, that we had to commit much to memory.

Not surprisingly, in an operation this memorable, completing this task wasn't as difficult as it might sound.

Dennis and I agreed on a schedule for the observation post: twenty-four hours on, then twenty-four hours off. The first eight hours of the off-twenty-four period were spent developing photos in the sixth-floor photo lab at the U.S. embassy. With each roll of development, we both became more and more proficient.

This OP reaped rewards from the very beginning. The two of us took no fewer than twenty rolls of 35 mm photos in the first weeks. We used ISO 200 speed film—the easiest to handle in the photo lab. Every time Carlos or his companions took a step outside that apartment, either Dennis or I captured them on film. We remained clandestine and undisturbed in our position. We were ghosts, shooting through the thin black curtains amid the fetid squalor of this sixth-floor apartment in K-Town. The work evoked a feeling of quiet power.

When I zeroed in on Carlos as he came to the gate of his compound, I often thought, "This guy is nothing but a drunken has-been, but he is dangerous." He drank almost every day, at places like the Greek Club and other spots where he would be sure no one knew who he was. I knew he was dangerous, though—he carried both a shoulder-bag weapon and an ankle-holstered pistol everywhere he went. He was no one to trifle with, but I sometimes found it difficult to bend my mind around the difference between what I saw through the Canon lens and the legend of the Jackal.

The presence of his bodyguards was a consideration at the beginning of this operation, but they were proving to be no factor at all. Tarek was the first, to be replaced over time by a series of men like him—broad-chested weightlifter types who stood at least six feet tall.

They were Arab, which meant they were trained in a specific way that made our job easier. During my twenty years in Arab countries, I knew their individual perspective and area of interest were different from those of an American. Arab bodyguards have a tendency to blast out in front of the person to be protected, similar to the manner in which some Secret Service handle the president of the United States. The Arabs concern themselves with a 360-degree line of sight.

Serious bodyguards, I have learned, do a lot of
surveillance before exiting a position. They view
the area from inside the house or building, taking into consideration every possible location for
an observation post. Well-trained bodyguards
also know that surveillance is normally performed from high positions, but Arabs, for
whatever reason, don't spend a lot of time looking up. This trait was consistent through numerous bodyguards, and it is a big mistake on their
part.

Carlos's men had a routine. The bodyguard
would exit Carlos's apartment, open the pedes-
trian gate, and walk to the east, toward Airport
Road. He would then walk around the west side
of the apartment. I suppose they were searching
for potential ambushes. When the bodyguard
was satisfied no danger lurked at ground level, he
would reenter the apartment. Soon after, out
would come the Jackal, sometimes with Lana
ASJ.

In fact, many times Carlos and Lana would
exit the apartment alone and depart, with Lana
driving and no bodyguard in sight. When I
watched this take place, I grew more convinced
at the ease with which we could have taken Carlos out of the picture and placed the blame else-
where.

Carlos lived hard. Despite his drunken altercation with the shopkeeper—the altercation that
led to his phone call to Tarek and our discovery
of him—he continued his habit of partying till

late in the evening. He would leave his apartment in the early evening, driven by either Lana ASJ or his bodyguard at the time. It sure seemed like a reckless lifestyle for a man attempting to protect his identity.

Carlos had a schedule—a loose schedule, but a schedule nonetheless. He was up at about 1000, and his usual routine was to come down to the courtyard, where the Cressida was parked, and spend a few moments looking out of the gate.

At about 1100, he and Lana would travel. Sometimes the bodyguard on duty would accompany them; I remember Carlos driving only once. He would go drink and return around 1600, then go out again at 2000 and come home for good between midnight and 0200.

Our sleeping and eating habits in the observation post were subhuman. I kept to Carlos's schedule, at least to some extent. I was awake long before he was, anticipating his arrival in the courtyard, and I usually went to sleep an hour or so after he completed his nighttime revelry.

As for eating, rice cooked on the burner in the apartment was the staple. I also ate all grades of military rations.

Creature comforts were nonexistent. On the same floor as our apartment was a small closet that served as a bathroom. Inside the closet was a "bombs away" spot that sat above a straight pipe that dropped down six floors. Into what, I didn't know and didn't want to know. When the situation presented itself, your job was to squat, squeeze, and hope your aim was accurate. It was truly awful in that respect, but it could be pissed in quite easily. We did have that going for us.

The stench made it impossible to breathe in the damned thing, and I imagine the landing zone must have been lovely. I adopted the practice of abstinence—waiting until my twenty-four hours were over, then using the toilets at the U.S. embassy, which were better geared for use by lower sections of a U.S. body.

During the third week of operating the OP, around June 16, we began to use the almost comically large 4,000 mm lens that was delivered via diplomatic pouch in a footlocker-type case. Accompanying this monster was an extra-rigid tripod that weighed roughly eighty-eight pounds by itself. During the hours of darkness on June 17, Dennis and I somehow managed to get the entire load up the steps of our dilapidated apartment building and set it up inside the OP.

I often thought, *If anyone from the outside world were to look inside this horrendous apartment and come across the collection of insanely expensive photography equipment, they would surely believe their eyes were lying.* As I surveyed the room, it was indeed a surreal sight.

This new lens was something else, period. It could be adjusted from 1,700 mm to 4,000 mm. It was fitted for a video camera, but we had a fitting that allowed it to adapt to a 35 mm still camera.

Our OP sat very close to the Khartoum International Airport, and after I set up the new lens I decided to see if it would allow me to monitor activity at the airport. I focused my eye through that lens and couldn't believe what I saw. First off, the lens saw right through the thin black curtains and up to two kilometers from our window. This power gave me a crystal-clear view of activity occurring on the airport grounds. I was eventually able to photograph all the security positions of the airfield and even identify the offloading of an Iranian cargo aircraft.

I reviewed our OP, knowing we had a fine setup for recording all arrivals and departures of Carlos and Company. We had a Canon 35 mm camera set up on a sturdy tripod with an 800 mm lens attached. We had The Big Lens on an even sturdier tripod, and another Canon OES camera with a 1,000 mm mirror lens attached. In the OP were two spotting scopes, binoculars, and all the good things it takes to conduct an operation of this sort. My partner and I existed in a state of constant excitement about this operation, for our target was one notorious man. The *most* notorious, in fact.

No one, but no one, knew what in the hell we were conducting here, but as a precaution we had to devise an escape plan in the event the building was raided. Knowing the sloppiness of the Sudanese military, we quickly came to the conclusion the police would announce their arrival by making a heck of a lot of noise. Stealth, it's safe to say, was not their trademark. I didn't sleep much at all, especially during this operation, and I knew I would immediately awaken if any *jundis* arrived in front of the building.

As little as we respected the *jundis,* we could not let hubris get the best of us. The reality was stark: If we were apprehended in the OP, we would most likely be shot on sight. We decided our quickest and surest escape route was out the back of the apartment building, straight down the outside wall via a rappelling rope. This would take us off the top of the roof and down six flights to Thirty-ninth Street. Dennis and I rigged the rope, attaching it to the sturdy pipes on the roof. This roof had a drop-down set of stairs that took some real strength to open, so the roof was nearly impossible to access. We left two pairs of rappelling gloves under an air vent, in case a future need arose.

Dennis and I calculated that we could notice the PSO charging up the steps and still get to the roof, rappel to the ground, and escape in the time allotted for such maneuvers. Of course, we would have to abandon our expensive—and cumbersome—gear. We took this plan to the COS, and he authorized its use in a worst-case scenario. Addressing this problem and coming to an agreement on a potential solution set my mind a bit more at ease.

We were in constant contact with Cofer Black, and during a chat in his office the COS said, "Billy, I want a face shot of Carlos, as close as you can bring that face into the camera."

"As close as I can bring him in, boss?"

"That's it, Billy. Close, as in C-L-O-S-E."

I rogered this order and proceeded to the observation post, where I set The Big Lens up to 3,200 mm, figuring 4,000 mm would be too close for any sort of clear photo. I attached the camera to the rear of the lens and let the camera sit there, directed at Carlos's apartment, awaiting his exit from the building.

It took about three hours, but *voilà*—Carlos appeared. He exited the door of the apartment building and stepped to his vehicle. I jumped across to The Big Lens and adjusted it for a perfectly clear shot of his head. I noticed something in his mouth as he opened the front gate slightly. He peered to the left, then the right, as I took a few shots with the "continuous fire" button switched on. After a few shots with the lens set at 3,200 mm, I flipped the lens wheel to 4,000 mm and fired again.

Should be great shots, I thought. The camera was absolutely still as I used the automatic-firing mechanism to take each photo. I took about ten frames of Carlos and watched as he turned and reentered the apartment building.

I was satisfied. I was convinced I had a fine shot of Carlos's head in the roll.

I was also pretty sure I had one the COS could not bitch about in terms of C-L-O-S-E-N-E-S-S.

The next day, after completing my twenty-four-hour shift and heading for the embassy to develop film, I was especially anxious to see the results of the large-lens photos.

I processed this roll and watched as the blank white eight-by-ten sheets of photo paper developed into great photos. As I dried the film, I was able to discern the object in Carlos's mouth: a toothpick. The photos taken with the 3,200 mm lens were just great, close up and clear on the target's bloated face.

When the 4,000 mm photos cleared, I saw a sight that made me laugh out loud in the darkroom. This amazing lens had brought the face in so close that all I had before me was a photo of his teeth, with the toothpick sticking out of it.

An eight-by-ten, extra-clear photo of Carlos the Jackal with a toothpick sticking jauntily out of his mouth.

Cofer would go bananas when he saw this.

I took these developed photos down to his office and knocked on his door.

"Come in, Billy, what do you have?"

I handed him the envelope with the photos inside. I was careful to arrange them with the photo of the teeth and toothpick on top of the stack.

I said nothing.

When Cofer pulled this photo out, he looked at the teeth and the toothpick. He pursed his lips and furrowed his brow. He turned it upside down, then sideways.

"Billy, what in hell is this supposed to be?"

"Boss man, you said C-L-O-S-E, and by God, that fucking photo of Carlos right before you is C-L-O-S-E."

He looked back at the photo. The realization

of what he was seeing dawned on him and he looked up at me. We stood there in his office, laughing at the absurdity of these photos.

He went through the rest of the photos, and the look of satisfaction was plain on his face. He nodded with each successive shot, perfectly framed and clear as a spring morning.

"Now, these are great," he said.

In the days that followed, Cofer made sure everyone in the station got a good look at what became known as The Teeth Photo. That photo landed on every single desk in the building, and everyone got a big kick out of the joke. There wasn't much levity in K-Town; when it came, it was welcome.

Independent contractors could not stay in Sudan indefinitely. I had to filter in and out, usually every six to eight weeks, and during the final week of June, an older SF retiree named Santos T. reported to K-Town to relieve me. One old SF man replacing another, meaning there were now three of us working this OP detail.

I spent slightly less than a month in the United States, chomping at the bit to get back to that disgusting building on Thirty-seventh Street. I kept tabs on the operation and counted down the days to my return.

When I did return, I relieved Dennis, which meant the post was manned by Estavan and me—two old SF men. An agency worker arrived at the Khartoum station to work full-time as a photo developer—a welcome addition.

Carlos the Jackal was the main target of this operation, of course, but Cofer Black and I were both interested in the people who visited the apartment building. My interest was piqued after watching many well-dressed males go in and out of the building. A good number of these I knew to be Iraqis, whose distinctive features can be identified by a good-quality photograph.

I was intrigued and hopeful we could nab someone like number two terrorist Abu Nidal, who was reported to frequent Khartoum. If we could double our pleasure by taking down two bad guys, why not? I began to prepare a whiteboard collage of all the visitors to the apartment building. I placed a photo of Carlos in the center of the collage and surrounded it concentrically with photos of each visitor. I noted the particulars—date, time, group—of the visitors, in case a pattern developed. Cofer Black briefed Ambassador Donald Petterson on this practice—the ambassador needed to be kept in the loop, but he didn't need to know everything.

In early July I noticed that an older local man began to make a habit of squatting on the corner of Airport Road and Thirty-fifth Street, adjacent to Carlos's apartment building. He would enter the apartment building in the morning, for varying periods of time, and then spend most of the day on the street corner. I surmised he was either a lookout for Carlos, or a lookout for the PSO. There was also a chance he was both.

Around this time, there was an even more perplexing development at Carlos's place of resi-

dence. A desert tent was set up in front of the
apartment building, occupied by Arab soldiers,
or *jundis*. Obviously, these guys were performing watch duty on Carlos's home. And just as
obviously, we filmed and recorded their presence, reporting our findings back to headquarters.

My collage was growing. Carlos's visitors surrounded him on the whiteboard. Those who
walked into the building, those who left or arrived with him in the Cressida—they all found a
place on my personal rogues gallery. I also began
sketching what I saw outside the OP window, in
perspective, and became pretty proficient at this
task. My view of our view, as it were, was excellent. The portability of the sketches became a
benefit during COS briefings with the ambassador or other visiting VIPs.

I hoped all this work was building toward a
conclusion, and in July there were signs of movement. The chief of station from Paris, a man
named Dick Holm, visited the Khartoum station.
Cofer Black introduced me to Holm as the man
who had taken the original photos of Carlos,
and that I working in the OP at the present
time.

Holm was distinguished by severe burns on his
face and arms. Black explained to me that his
friend had been burned during an action in the
Congo, where the two men had worked together.

The two chiefs briefed me on the political as-
pects of the Carlos mission. Since the United

CONCLUSION OF THE HUNT FOR "CARLOS THE JACKAL" –
IN KHARTOUM SUDAN – 02/08/1994

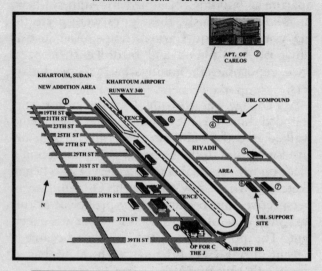

▰▰▰ = STREETS	③ **U.S. MANNED OP V. C THE J**
▬▬ = U.S. POSITIONS	④ **RESIDENCE OF UBL 1992/93**
▬▬▬ ADVERSARIAL SITES	⑤ **U.S. RESIDENCE KHARTOUM**
① **IBN KALDOUN HOSPITAL**	⑥ **UBL LEADS PRAYER EA DAY**
② **LOCATION OF C.'S APARTMENT**	⑦ **PHOTOS OF UBL FROM HERE** NOT TO SCALE

<u>SUMMARY:</u> C THE J WAS THE SUBJECT OF INTENSE SEARCH IN KHARTOUM, SUDAN, FROM DEC 1993 – FEB 1994.

<u>02/08/94</u> -THE VEHICLE, USED BY THE BODYGUARD OF CARLOS, WAS SIGHTED AT THE IBN KALDOUN HOSP, ON 19TH ST, IN THE NEW ADDITION AREA OF KHARTOUM CITY, ON 08 FEB 93.

-THE AUTHOR AND ONE OTHER IMMEDIATELY DROVE TO A POSITION TO PHOTO ANY AND ALL WHO APPROACHED THE VEHICLE. A LARGE CAUCASIAN MALE (ID OF THE MALE WAS MADE 48 HOURS LATER), EXITED THE HOSPITAL AT 1500, 08 FEB 94, ACCOMPANIED BY AN ARAB FEMALE, WHO THE AUTHOR KNEW WAS THE WIFE OF C. A DIVERSION DREW THE ATTENTION OF THE MALE, WHILE 31 PHOTOS WERE TAKEN AND SUBSEQUENTLY DEVELOPED AT THE U.S. EMBASSY IN KHARTOUM

-PHOTOS WERE TAKEN BY THE AUTHOR, AND VERIFIED IN JORDAN (THREE DAYS LATER) AS BEING ILICH RAMIREZ SANCHEZ "CARLOS THE JACKAL." THE HUNT HAD ENDED, SURVEILLANCE THEN BEGAN.

<u>30 MAY–AUG 1994</u> -OP SET UP IN A CONVERTED HOSPITAL FROM MAY–AUG 1994. PHOTOS OF CARLOS, PLUS ALL WHO VISITED CARLOS, WERE TAKEN, THEN DEVELOPED.

States had no active warrant against the Jackal, we could not be the ones who took him into custody. We would continue the OP overwatch, but the plan called for the Sudanese (we hoped) to turn Carlos over to the French Direction del la Surveillance du Territorie (DST) in Khartoum. This group, the French version of the FBI, would then transport Carlos to Paris.

The head of the DST, Philippe Rondot, visited Khartoum to meet with Cofer Black during the last part of July. Rondot had a major vendetta against Carlos stemming back to Carlos's murder of two unarmed DST officers who visited Carlos's apartment in Paris. Rondot's history with Carlos traced back to the 1970s, when he tracked him across a good portion of the globe, from Algiers, Colombia, Cyprus, and Greece. We were attempting to get the French interested in taking down the Jackal. We had identified Carlos and could have apprehended him in a second, but the French had the best case against him. This seemed like the quickest way to get the fat-ass celebrity terrorist off the street for good.

After Rondot arrived at the Khartoum Hilton, I was to tail him as he walked to a meeting in the hotel with Cofer Black. I was providing cover for this meeting—ensuring Rondot was not being tailed by another person. I was behind him for about ten minutes before I realized he was lost. When he made a wrong turn on a lower floor of the hotel, I grabbed him from behind by the belt, in a way that prevented him from turning toward me.

I said, *"Suivez moi."*

Rondot was a big man, six feet, three or four inches compared with my five feet nine inches, but this unexpected encounter scared the living shit out of the Frenchman. He got where he needed to go, however.

We weren't necessarily turning Carlos over to the French at this point. We were just getting acquainted, letting them know what we'd done and the evidence we had. We were ready to let Rondot know that we knew where Carlos was, but we weren't ready to share the exact location. It was our work, and it was still our little secret.

The plan was for the French to get the Sudanese government interested in turning Carlos over. I wasn't privy to everything, but I did know that some of our photos were being shared with Rondot. He, in turn, was to confront Hassan al-Turabi—the government strongman—with the photos that would prove that Carlos was, in fact, holed up in K-Town and had been for months. France was prepared to withdraw all aid to the country and make statements to the world revealing that the Sudanese government was harboring the notorious Carlos the Jackal.

Rondot came and Rondot left. If negotiations with the Sudanese did indeed take place, the results were not made available to me and the other ICs. At the beginning of August, I made a quick trip to the United States and returned within a week. On the afternoon of August 12, 1994, sitting in the fetid OP, I watched Carlos depart his residence with Lana ASJ at the wheel of the Cressida.

I reported this fact on the commo net and sat back, fully expecting the couple to return later in the day. I waited through the afternoon and into the evening, but they did not return. I reported this fact, via encrypted portable radio, to the COS.

Cofer replied, "Give me a call when they return, Batman."

"Roger."

I remained vigilant during the night, but the situation remained the same. No Carlos. No Lana ASJ. You know what I was thinking—*Goddamnit, we let these two slip through our fingers*. I was worried he had received word the noose was tightening around him, and my worry was exacerbated by one fact: Carlos and his bride had not remained away from their apartment for an entire night since the OP had been established.

Throughout the morning hours of August 13 . . . nothing.

By this point, I would have bet money they had fled Sudan.

I was distraught.

All this work for nothing.

At precisely noon, August 13, 1994, as I sat dejectedly in the OP waiting for Carlos to return, the COS called me via encrypted radio, telling me to come to his office in the embassy.

"Cofer, then no one will be in the OP," I said.

He replied, "Lock the place up, Billy, and come see me."

Donald Petterson, the ambassador, stood near the Marine guard outside the embassy when I arrived.

"Hello, Billy," he said with a big smile. "Great work."

This remark came totally out of the blue. I had no idea what he was talking about.

Inside the station, I opened the cyber lock and entered.

Immediately, Cofer and the fine lady station manager handed me a glass of champagne.

Cofer bellowed, "Toast, Billy, you sweet son of a bitch. Carlos is in prison in France."

My emotions did an abrupt about-face. My frustration and anger were immediately replaced by a wave of euphoria. I thought, *Goddamned if we didn't get the job done, after all.*

We celebrated a bit, pretty mildly considering the circumstances. As always, though, there was work to do. We got busy with our debriefings. I asked Cofer when he wished to close down the OP, and he said, "Let's wait a week, so our closing will not be correlated with Carlos's departure."

I went down to the embassy snack bar, where AFN TV was showing Carlos the Jackal being transported in Paris, France. The French aircraft was landing at the military airport, with Carlos in tow.

The announcer on CNN was saying, "And the

French have searched out and found Carlos the Jackal, in Khartoum, Sudan. They have conducted an apprehension operation in that city, taking the most dangerous terrorist in the world, alive."

"What bullshit," I said aloud to the snack-bar television set.

But at least the son of a bitch was out of business.

That night Estavan was in the OP, which was to remain operating for one week. I decided to jog at 0300—my usual jogging hour. The *jundi* curfew point was abandoned at this hour, so I had the run of the city. As I jogged around the Riyadh Road area, toward Airport Road, I saw traffic in front of Carlos's apartment building.

I saw several station wagons and a large truck parked outside the building, their engines running, their lights on.

I saw about twenty armed *jundis* forming a perimeter around the building.

I also noticed a large and very black man with a *jalabeea*—turban and white robes—atop his head. It appeared to me to be Sudanese vice president Hassan al-Turabi himself, so I stopped and took cover.

I called Estavan in the OP with the portable radio I always carried with me.

"Are you watching this scene below?" I whispered.

"No," he said a little groggily.

I asked him to get some photos of the scene

with the night lens. Estavan completed the mission with great care and excellence. The high-speed film revealed the large black man was, in fact, Hassan al-Turabi, who was overseeing the removal of Carlos the Jackal's property.

From my position, in a row of bushes no more than fifty meters from the building, I saw Lana ASJ exit the building, wearing slacks. She was crying and carrying a suitcase. She was led to a police vehicle and driven away. It appeared all of the couple's gear and furniture had been loaded in the truck and vehicles. Before dawn, the area was clear.

At 0800 I reported this information to the COS, who said to me, "Billy, what in the hell are you doing out there at 0300?"

"Keeping fit, boss."

Since my early morning work had reaped good, solid information, Cofer said nothing else to me. The photos developed beautifully, and we were armed with further proof that Hassan al-Turabi was deeply involved with Carlos the Jackal.

Within a week of Carlos's capture, Estavan and I removed all the gear in the OP, secretly and piece by piece to keep from alerting the neighborhood that we were departing. The lenses, the cameras, the Bunsen burner, my whiteboard collage—all of it went out of the disgusting apartment and down six flights of stairs.

A few days before that, on the third day after the reporting of Carlos the Jackal's capture, I looked down from the OP and noticed another large-scale commotion in front of Carlos's former residence. I saw many vehicles and gear belonging to several French television stations. Producers and reporters and gofers were in and around this apartment building like a plague of locusts, frantically hurrying to report this magnificent French coup.

CHAPTER 12

It's standard practice for a person in my position to recount not only where he was on September 11, 2001, but what he felt and what he believed as he watched the World Trade Center towers explode and fall. And then, given my extensive work in the area of counterterrorism, I'm expected to put forth a theory on the intelligence breakdown and identify the silver bullet that could have prevented the attacks. Well, I can't do that. I know that we understood the threat bin Laden posed, and the time I spent watching him in 1991 and 1992 is the evidence that we identified the threat early. I might have been slightly less surprised by the terrorists' capacity to inflict this kind of pain and suffering, but I knew their brand of hate allowed for such atrocities. In all, my gut reaction—shock and revulsion—was the same as your gut reaction. Simply put, the previously unthinkable became real.

I was doing some contract work for the CIA

and happened to be in the Langley headquarters building when the attacks took place. Within minutes, warning alarms sounded within the building. We had practice alarms often, to ensure that everyone understood their particular evacuation routes. This, however, was the first time the alarms sounded for real. We had witnessed the hits on the World Trade Center and knew the hijacked United Airlines Flight 93 remained unaccounted for. It was a widespread assumption within the building that this flight was headed straight for us in the CIA headquarters. There was no panic, just an understanding that those in my division needed to walk to the west parking lot, away from the buildings, and await the inevitable impact.

Once outside the New Headquarters Building, standing in the parking lot, several of us listened to the sequence of events through my portable AM/FM radio. The parking lot emptied pretty damned fast after word circulated that we were to get off the grounds ASAP. Highway 123, which runs east and west just north of headquarters, was jammed with vehicles that had departed downtown Washington, D.C., and headed for northern Virginia and western Maryland.

Upon hearing that Flight 93 had gone down in a Pennsylvania field, a couple of us returned to the HQ building to pick up any necessary gear. I headed to my apartment, down the George Washington Parkway, but I could not get into the Arlington area. I doubled back to Route 66 eastbound and managed to get past the Pentagon, where I watched the building burn while the police and highway patrol set up a roadblock to keep all traffic out of Arlington. Eventually, I managed to return to my temporary home.

As the days passed, my fellow workers and I were aghast to realize these al Qaeda operatives, with a lot of direction from higher up their chain of command, had successfully completed an operation this complicated and bold.

We in the agency knew what the world quickly discovered: These nineteen al Qaeda hijackers had not only pulled off these simultaneous terror operations, but they had trained to fly commercial aircraft right under our noses, on U.S. soil.

I had seen this anti-American movement firsthand, up close, face-to-face. Beginning in K-Town in 1991, I saw how these people sat in stony-eyed worship in the presence of Usama bin Laden. I realized this tall, bearded, and slightly bent younger man who led the 1210 prayer each day had a special hold on his people. Through the lens of my camera and its special magnifiers, I watched the workers in attendance lean forward in their squatting positions to hear the quiet words of this man they considered their messiah. I couldn't hear his words, but I read the body language of his followers and clearly understood the depth of their convictions.

And still, on the morning of September 11, I was stunned.

Stunned and angry. And determined to place myself in a position to do something about it. I was seventy-one years old when those Towers fell, but I didn't give a good goddamn how old I was. My country was going to war, and for nearly fifty years that meant only one thing: I needed to get myself into this war also. Whatever war we decided, I needed to be involved. This gangly bastard had engineered a stupendous blow to America, witnessed by millions in our own homes. By God, this man and his al Qaeda punks needed to be killed and tossed into the garbage dumps of wherever they may be found.

My quest for adventure had not wavered. I did not wish to remain on the sidelines reading about actions conducted by younger, stronger men and women. My craving is, and always has been, to be involved in actions conducted to ensure America remains strong, safe, and free of those who have its destruction as their goal.

I was more than ready to argue, connive, and cajole for the opportunity to join this battle. At my age, I knew it didn't look promising, but there was always hope. After all, I'd been faced with the same kind of odds before. I'd overcome them then, so why not now?

Shortly after September 11, President George W. Bush issued his declaration of war against all terrorists and fingered Afghanistan's Taliban regime as a central culprit in the harboring of terrorist cells. Afghanistan was known as a hotbed of terrorist activity, similar to Sudan in the early to mid 1990s, and bin Laden's al Qaeda was prominent among those who used the sanctuary of Afghanistan to train for their particular brand of war. The target was obvious; preparation for war in Afghanistan began September 12.

The geocoordinates didn't appeal to me—I liked my wars muggy and hot—but if Afghanistan was the chosen theater, my sissy Texas ass would have to find a way to keep warm. I am a devoted "10/10 man," meaning I wish to fight my wars in territory between 10° north or 10° degrees south, preferably at sea level. I knew Afghanistan was damned sure not in the 10/10 zone, but I was determined to get over this and not be bothered by it. Age be damned, I began working diligently to place myself on the field of battle. This effort—my own war within a war—began nine days after the World Trade Center destruction. I approached my superiors to let them know I was ready, willing, and—most important—able to be a valuable asset to our quest to topple the Taliban and ferret out the al Qaeda operatives who made Afghanistan their home.

Taliban leader Mullah Mohammid Omar took a calculated risk by refusing the Bush administration's offer to surrender UBL and his underlings. Perhaps Omar felt immune, believing the United States did not have the stomach for a war so far from our mainland. Perhaps he simply recalled his Taliban forces defeating the Russians and believed the same fate would befall the Americans. Whatever the case, it was a horrible miscalculation on Omar's part, because the American people were united to root out those who had a hand in killing close to three thousand people in New York, Washington, D.C., and Pennsylvania.

My personal mission was similar to my effort to return to Vietnam following the horrible combat wounds I suffered in Bong Son. I needed to put my powers of persuasion to work in new and creative ways, and I needed to do it pronto.

I will not reveal my department or the particulars of its hierarchy, but I knew some of the bosses quite well. I had worked with them in Vietnam and afterward, and many of them knew of the wonderful success of the Carlos operation. This put me in pretty good stead with these men—good enough to give me access to direct talks in their inner offices.

I have to admit, the topic of my sanity was a recurring theme in these discussions. The initial reactions to my request typically went something like this:

"Are you crazy, Waugh? You're seventy-one years old."

Or "Do you know how goddamn cold it's going to be over there, Billy? Are you serious? Are you crazy?"

Was I? I don't know, but if having the guts and ability to track down some bastard who intends to bring harm to the United States is crazy, then I guess I plead guilty as charged.

Day after day, I worked to convince several of these men that I would be an asset in the battle area of Afghanistan. I emphasized my thorough knowledge of search and rescue. I had repeatedly demonstrated my understanding of war and the weapons of war. On top of everything, I convinced them I was in the kind of superior physical condition—no matter the age—that would enable me to withstand the rigors of both weather and war in this remote, frigid, and elevated part of the world.

I called in all my favors and summoned all my powers of persuasion. This wasn't just a vanity mission; I was convinced I could help and felt the need to convince others, too. In the process, I guess I made a nuisance of myself. I'm not too proud to admit that.

Finally, after several days of this back-and-forth, when there was really nothing left to say on either side, the boss told me, "OK, Billy. Goddamnit, go on and get out of my hair."

I was thrilled to hear this. I said, "You will not be sorry, boss. I will do well."

"I know it, Billy," he said. "Now scram."

As I prepared for my final combat mission, I contemplated the course of history as it pertained to the years I spent serving my country. The overarching enemy of our way of life had changed from godless Communism to religious zealotry, and I saw both battles from the ground up. From my perspective, the new threat was much more insidious and threatening. In my estimation, terrorists are 100 percent more difficult to find, manage, and fix. They are spectral and shadowy, planning and acting in small groups.

When all angles are considered, it becomes obvious that small-cell terrorist groups are the only method these enemies have to do battle against a magnificent military such as ours. Our might is unprecedented and unbelievable. Our ears and eyes are everywhere except inside the homes, minds, and hearts of the individual terrorists.

The way it looks now, Saddam Hussein spent millions on a military but resorted to terrorist tactics when it became obvious his army was no match for the might of the coalition forces. Since the official end of the war the isolated and individual acts of terror have proven to be a major disruption for our side. Had he adopted this position from the outset of the war, and one hundred or so of the thousands of fanatics had succeeded, our military power would have been completely overshadowed. The media would have ensured our departure or failure with their reporting on the devastating impact of the terror campaign.

I knew my final combat mission would be nothing like those that came before. The terrorists' use of religious justification for their actions intensifies the problem of changing and fixing them. For the better part of twenty-four years, beginning with my extensive work in Libya from 1977 to 1979, I had become fascinated—and often frustrated—with individuals in the Arab world.

Most of the Islamic people I encountered in countries such as Sudan, Iraq, Egypt, Libya, Yemen, Syria, Jordan, Turkey, Uzbekistan, Albania, Macedonia, and Saudi Arabia were reasonable in their practice of religion. To understand the culture, I familiarized myself with the Koran enough to understand the book is rife with wonderful words of brotherhood and cooperation among Islamic mankind.

Arab families were as tight-knit as I had ever seen. The Arab children listened to their extended-family elders and heeded the words of their parents. In the Libya I experienced, an Arab would never steal, rape, or partake in other forms of criminal activities, for these men and women believed the words in the Koran were sacred and required absolute fealty. In fact, I don't believe the Arabs who stole and looted after the Battle of Baghdad considered this stealing at all. I believe they committed these offenses to gouge Saddam for spending their wealth on lavish dwellings for himself and those around him.

However, I was also well aware that some sects of Islam wish infidels destroyed. Infidels, in this case, means any non-Islam. These sects blame every ill of their world on Western infidels. They hate the Israelis with a passion beyond reason, of course, and deem it an honor to explode themselves among Israelis of any age or sex. These acts are applauded by most of the Islamic world. They are, in effect, the Islamic answer to our smart bombs. Our precision strikes seek out a specific group, vehicle, or building at a certain geographical spot on earth. The terrorist with hidden explosives—an individual "smart bomb"—seeks out a group at random, but the precise nature of the attack is impressive and effective.

As I waited for the call to the war zone, I thought about the nature of this new enemy. The depth of a belief that calls for martyrdom for a religious cause, misguided as it might be, is completely foreign to our Western minds. We have no comparable passion in our culture. Though our attitudes are gradually changing, we still envision war being fought on a battlefield, with armed men lined up versus the armed enemy, to plod forward on command to take over the enemy's territory and claim it. Tanks versus tanks, men versus men, artillery pieces versus artillery pieces.

The United States threw this equation completely out of balance with the introduction of pinpoint bombs and missiles that come in from the sky and travel unimpeded to their exact destinations. The Arabs have countered our technology with human bombs that travel (often impeded) to their destinations. The Arabs who are able to circumvent the many barriers to complete their "bombing run" are considered by many to be national heroes.

Our bombs/missiles travel a limited range from the mother ship to the destination; whereas, the Arab bomber must travel through all sorts of human-erected barriers to reach the intended target. If the terrorist sects that advocate the launching of individual human "smart bombs" were better organized, we would be in one hell of a lot more trouble than we are in now.

These loosely organized groups—led but not

dominated by al Qaeda—espouse terror and destruction as a tool for moving Islam to the top of the religious powers. I have spoken to some of these terrorists, and they consider terror attacks against the general public their only outlet to hurt and destroy the infidels who have wrongfully ousted them from their homes so many years in the past. I worked right there with these al Qaeda operatives and heard these arguments firsthand many times, especially during an assignment in Yemen. Under no circumstances would I enter into conversations concerning the rights and wrongs of U.S. policy. I listened to many Arabs condemn the United States for backing Israel or some other policy, but I would never express my opinion. There were times—especially in Libya in the late 1970s—when I felt like choking a couple of the Arab officers. They were about as informed on world politics as the goats they ate for dinner each day. But my job was to teach these yahoos a few uncomplicated things, and to gather information. I did this, alone and single-handedly, and I stayed out of the debates. My typical response was a simple grunt or an empty word or two intended to defuse the zealous type.

My experience taught me that these al Qaeda "warriors" are not concerned about the effect these indiscriminate acts have on the innocent. They view this technique as the sole manner in which they can strike at their declared enemies.

Terror brings notoriety, and the more gruesome the act, the greater the notoriety. Notoriety, in turn, brings money in the form of donations from sympathetic followers and governments. This is the self-fulfilling nature of terror—each act of terror brings about a greater ability to conduct further acts in the future. These groups owe their survival to their ability to inflict death.

I do not profess to be an authority on Arab beliefs and views, but I have lived among Arab military men and had close contact with them for many years. I became intimately aware of their views through my position as teacher of military tactics, weapons, and basic techniques of warfare. Arab students such as the ones I taught during my stay in Libya resented the fact that I, as an American, stood before them as their teacher. I also taught many Arabs who were in the intelligence field in various nations—Yemen, Jordan, and others. In their eyes, I represented the enemy.

They were not afraid to reveal their thoughts concerning the Israelis, and the U.S. sponsorship of the Israelis. I listened politely but assumed a position of aggressive neutrality in my day-to-day dealings with these men and their officers. Politics was not my job, and I made sure each one of them knew that from Day One.

Before each of my teaching sessions, I stood before the class daily declaring my intention to remain out of the political and religious fray.

"I am not concerned with politics," I would tell them, "or any of the vagaries and notions of Islam versus Christianity. Period. I am here to teach you this weapon/compass/map, or these methods of maneuver. I will not participate in any discussion of a political nature."

Usually, this rehearsed statement was enough to keep religion and politics—if they can be separated—at bay. However, many times an individual Arab would attempt to draw me into political discussions. Once again, I used the opportunity to repeat my statement. I would not reveal my own political views. If the person or persons persisted in baiting me, I would simply walk away.

Those experiences made me aware that the Arab military sympathized with nonsecular terrorism as it was employed by the likes of al Qaeda or, to a lesser extent, those who employed Carlos the Jackal. Terror was their trump card, and September 11 was their best hand yet.

Now I found myself in yet another position to hasten their demise. If the mission of these terrorists is to die, then our mission was to accommodate their wishes. Terrorists truly believed a death at the hands of the Americans—providing they took some of us with them—would secure their place next to Allah, at the side of his disciple Mohammed. Our side wanted to assist them in this goal without going along for the ride.

After being tossed out of my boss's office with the OK to travel to Afghanistan, I had no second thoughts and no regrets. I was disappointed with the pace, however—I wanted to get there in the first wave of Special Forces/CIA teams that landed in Afghanistan, but I was forced to wait nearly a month for travel. I didn't like losing a month, but my disappointment was tempered by the reality of the situation: I was going to war. My body hummed with a feeling I hadn't had for years. I was alive, ready to go.

During the second week of November, a group of us—I must remain vague regarding number and identity—boarded a USAF C-17 transport plane for travel to Germany, then on to Uzbekistan, and finally to Afghanistan. A long way to travel to the battlefield, and a long time to reflect on the number of times I had placed myself in a position to join the fray.

A sane man would not go out of his way to join battle, at least not more than a dozen times. However, entering this war was like winning a victory to me. I had been able to throw my hat into the ring, and I knew I could handle whatever was inside there. That ring had been my home so many times before. It felt comfortable and familiar.

I felt absolutely no trepidation or anxiety. The fear of entering battle had long since departed my psyche. Most men, I postulate, fear the unknown. As I sat on that C-17, I reveled in the unknown, anxious to take a winning plan into battle against a shadowy enemy. My attitude had never changed: If a man prepares for battle, trains for battle, studies the enemy, and practices for every possibility, the outcome of the battle will take care of itself. Fear has no place in the mind of the well-prepared soldier.

This impressive C-17 monster caused me to

remember my first trip on a USAF aircraft, a DC-3 also known as a C-47. It was 1948, fifty-three years previous, and eighteen of us fledgling, wannabe paratroopers prepared to make our first parachute jump at the Fort Benning, Georgia, Parachute School drop zone. It was my first trip in an airplane, and I had to jump out of the damned thing and land near the Chattahoochee River outside of Columbus, Georgia. Jump school was five static-line jumps, so it took me five airplane rides before I was able to experience a landing.

My mood was uncharacteristically contemplative as we headed toward the war zone. I knew this would be my final combat mission, and I knew it would be unlike any other mission that preceded it. I saw combat in the first Gulf War in a very low-profile capacity. I saw combat in Bosnia and Kosovo, conducting operations I cannot discuss. This war, however, would bear little resemblance to conventional combat. In the past decade, the technology of war had changed the battlefield. The days of emergency panels, mirrors, and randomly delivered dumb bombs were long gone. The days of calling for HE bombs and then hoping they didn't land on your head were long gone as well. We were working with the very smartest of bombs, and with the lads who knew exactly how they were to be delivered on the exact target where they were needed. I couldn't wait.

The way I saw it, I was not only going to war. I was going to school.

* * *

By the time we took the field, just after Thanksgiving Day, the Taliban and al Qaeda were in disarray and confusion. Our mission in southeastern Afghanistan was not to battle tanks and enemy ground units. Instead, we were tasked to ferret out these Talis, al Qaeda, and Chechens from the caves, tunnels, and safe areas south and southeast of Kabul.

My first stop was a newly captured airfield in Baghram, Afghanistan. From there, I was transported to a recently captured hotel in downtown Kabul. Over the course of the next week, I joined with a small CIA group that would move to the war zone in Lowgar Province, southeast of Kabul. Operation Enduring Freedom marked an unprecedented cooperative effort between Special Forces and the agency. The mission was to join with the thirteen-man Special Forces ODA 594 to create Team Romeo, which would continue the task of clearing the country of Taliban and al Qaeda. I was an independent contractor working with this team.

I couldn't hide my age, of course—I was 71 years old, and the next oldest soldier in this theater was 54. This meant I was something of a curiosity, and I came prepared for whatever reaction my presence would create.

There wasn't much waiting involved. The reaction began shortly after I landed in Kabul, when I was approached by a tall American who wore native Afghan clothing and looked as scrungy as I did. I didn't know the man, but he went practically buggy when he saw me.

"Aren't you Billy Waugh?" he asked.

I informed him that I indeed was Billy Waugh, and he immediately turned to the man next to him and said, "Let's get a photo with Billy."

I complied, smiling for the camera, then politely asked the men who they were. I discovered the tall man was none other than the commander of the 5th Special Forces (A), Colonel John Mulholland. His partner was his executive officer, Lieutenant Colonel John Haas.

"I know all about you, Billy," Mulholland said. He told me a few of the things he knew— good things, of course—as I grew more embarrassed and proud with every word he uttered. Mulholland and I met several times in Afghanistan, and we always spoke about the manner of the war and the morale of the troops.

Since Special Forces was my heart and soul, seeing such a fine leader as Colonel Mulholland in charge of these 5th Special Forces men made me feel proud. Of course, the 5th Special Forces was the main SF unit of the Vietnam War, and even SOG's records and assignments were to the 5th Special Forces (A). I knew at that very instant the Taliban was having its scruffy, disheveled asses whipped outright by the teams of the 5th SF(A). Damn, I couldn't wait to get right out there to join my old outfit.

I was determined to be more than a relic. My sales pitch centered on my ability to work as a liaison between the CIA and Special Forces groups, but I wasn't in Afghanistan to play diplomat. I prepared to join teams with the Special Forces ODA by familiarizing myself with several new weapons, especially the wonderful M-4 Carbine. I also re-familiarized myself with the AK-47 Kalashnikov rifle, used by both sides in Vietnam, and the AK-74, a lighter version of the AK-47.

I also took quick lessons and practiced with the latest fragmentation grenades. This grenade was larger than the minis I used in the Vietnam war, and even more lethal. I've always been an aficionado of grenades, while many of my comrades have been enamored of pistols. The pistol fascination never gripped me. I once attended a pistol-shooting course at a school run by a world champion shot.

"You know, Colonel," I told him, "I really don't worry too much about pistols and all of this fast-draw shit, for where I'm used to working, I usually carry a submachine gun and a ton of hand grenades."

He looked at me kind of funny, so I shrugged my shoulders and said, "I'll go for the masses with my grenades and fuck the pistol. But if it comes to me having to kill with this pistol, I guarantee I'll win that battle, too."

It wasn't just the weapons and munitions that astounded me. The new communications gear blew me away. The radios no longer operate ground-to-ground, instead bouncing off the Milstar Satellites orbiting 22,000 nautical miles above the earth to provide encrypted communications 24/7. This satellite alone cost $800 million, and the Titan launcher that transports it to the heavens comes in at another $1 billion or so.

No, we weren't in Vietnam anymore.

My personal gear was composed of the following:

- One -25°F sleeping bag; long drawers; numerous pairs of cold-weather socks; a Vortex jacket; one scarf; one Afghan headpiece called a *chitrali* (I knew for a fact the weather would be colder than cold)

- One AK-47 with seven magazines of 7.62 ammo; grenades with an H&K 40 mm grenade launcher

- One rucksack, with a large and small Camel-Bak water container

- Snap links

- One AN/PRC 112 survival radio; night vision device (NVD); two cameras; Garmin GPS

- One Leatherman knife; one Oldtimer knife; a compass; a mirror (still, and always)

- A few thousand dollars of personal money

We were lashed, loaded, and ready to go. It was -5°C on December 1—my seventy-second birthday—when we grabbed a few cases of

Meals Ready to Eat (MREs) and departed Kabul by trucks with a few dozen Afghan Anti-Taliban Forces (ATF). We were to pick up (read: hire) about one thousand troops in Lowgar Province, where ODA 594 would rendezvous with us.

When I joined Team Romeo, in Lowgar Province, my presence created another ripple of reaction from the thirteen Special Forces ODA team members. I say this reluctantly and without arrogance: These men practically worshiped me. This was somewhat embarrassing, but I attempted to roll with it as best I could. A seventy-two-year-old Special Forces veteran with eight Purple Hearts and combat experience tracing back to Korea was bound to create a stir among these men, and I prepared myself for the inevitable questions and quizzical looks. I figured these young men were thinking, "This old bastard must be nuts."

Their interest went beyond curiosity, though. Students of combat and aficionados of clandestine operations, the members of ODA 594 had read author John Plaster's books on SOG and had heard many of the unpublished stories of SOG and Vietnam. Some of them knew a sketchy outline of my more noteworthy work as an IC with the agency.

When the praise and adulation got too thick, I deflected the attention by bringing them back to the present situation. I warded off the praise with my walk-on-water line, and also by telling these young, bright SF men the truth: They were stronger and brighter than I had ever been, and they had the kind of equipment we of the Vietnam era could not even imagine.

However, every day I was reminded of my age, and every day I imagined someone giving CIA director George Tenet the news that Billy Waugh had suffered a combat wound while freezing his bony ass off in the desolate Afghan wilderness.

The scene, as I imagined it, would conclude with a wide-eyed, gape-mouthed Tenet saying, "What's that crazy old bastard doing over there, anyway?"

Upon arrival in Lowgar, our agency team occupied an old school compound. The children of Afghanistan, especially in remote areas, do not go to school during the cold and dangerous winter months. We set up camp in this tiny school area and vowed to leave it in much better condition than we found it. (We did this, too, and compensated those who ran the school with more than adequate rent.) This was a poor remote area, and therefore it was a poor, run-down school. There were only benches for seats, and the blackboards were dilapidated.

We acquired several kerosene-powered stoves and bought a fifty-five-gallon drum of fuel to power these heaters. Not exactly an ass-kicker, but if you placed your hands practically on the stove, it would do the job. With the basics in place, we established communication and defenses. Emergency combat plans were prepared. Language proved to be a thorny issue—only Farsi, Pashtu, and Dari were spoken in this area.

Arabic, the language of the enemy, was not spoken, rendering my limited skills meaningless.

We were met during that first week by a self-styled general with roughly three hundred troops. His name was Zaidullah, and he was the usual stripe of foreign indigenous leader—as unctuous as humanly possible. In my opinion he was a no-good cheating shithead who was not to be trusted under any circumstances. He didn't care about the war, his troops, or any of the Americans in his presence. His concern was lining his pockets with silver and gold. I had dealt with hundreds of these sorts, and it struck me every time how similar they were in manner and bearing. I was not the CIA team leader, so technically I was not negotiating with the local generals, but I did give advice at times when it appeared the team leader was receiving the stick's short end.

The Special Forces ODA 594 joined us during the first week of December. It was a fine team, led by Master Sergeant Darren Crowder, the team sergeant, and team leader Captain Glenn Thomas, who didn't have much experience but was wise enough to listen to his men.

The cooperation between the agency and the Special Forces ODA worked like this: The ODA trained the Pastun troops, who were often hired on the spot. Our small agency team worked near and around the ODA team. This was a new form of cooperation between these two separate outfits with separate chains of command. Sometimes the agency would be anxious to complete a job while the ODA needed to wait for permission to join in the fun. Given the circumstances, we worked well, and I considered it part of my role as a former SF man to smooth the waters whenever they got choppy. When you're heading into combat, with lives on the line, you can't just cowboy around and act indiscriminately. You have to have a decent plan, including air support, and when an action involves two distinct groups with two distinct sets of procedures, it can get complicated.

Master Sergeant Crowder was a no-nonsense type who took BS from absolutely no one. He and the agency team leader had a few frustrating moments, but it was quickly smoothed over. We all understood the game here, and we worked at keeping the peace. I continually used my status as the old combat sage to keep the interaction between the teams mostly friendly and cooperative. I often thought to myself, "Billy, this is not Vietnam, you must seek compromise." In Vietnam, at least in my recon companies of CCS and CCN, there was no compromise, just get out there and do it. I attempted to keep my finger on the combined pulses of the SF and the agency lads, to ensure tranquillity and equanimity prevailed.

I did not want to piss off either side, especially the SF side. Usually the friction was not overwhelming, but there were times interdiction was necessary. This was one of my selling points to my superiors when I was successfully coaxing them into allowing me to be here in the first place. One thing was for sure: I was able to work very closely with the Special Forces team, for I am a Special Forces man first and foremost.

It did my heart good to see these Special Forces men working hard and without complaint. Took me back a bit, I admit. It is imperative that a Special Forces man be trained to work and think swiftly. He is alone in the sphere of enemy influence, and fast decision making—plus continuous planning—allow the SF team to progress smoothly with the indigenous personnel the team must lead in combat operations. My sixty-two-day experience in Ba Kev was an extreme example of the thinking and planning that must take place when an SF team lives and works so many kilometers behind enemy lines, attempting to win the hearts and minds of an indigenous force.

I didn't make a habit of speaking of my Vietnam experiences while in Afghanistan. Many of the ODA 594 took part in Desert Storm, and those vets had briefed the noncombatants of the ODA on the perils of war. Vietnam was 180 degrees different from the war in Afghanistan.

We were called upon to fight a "standoff" war in Afghanistan, which means you get yourself above the enemy with small teams, and bomb him out of his jockstrap in order to kill him. Jungle warfare, of course, was totally different. In the jungle you must close with the enemy to kill him. For that reason I didn't preach to these men, although I did speak to them about the necessity of surprise and diversionary tactics to confuse the enemy. As I watched these men of ODA 594 prepare for battle, I knew they would excel.

Our group also included two members of the U.S. Air Force TACPs, whose specialties included the calling of Close Air Support (CAS) against the Taliban and al Qaeda forces. As expected, these lads were well trained and anxious to engage the enemy. To give you an idea of the strength of these Special Forces men, consider this: Each man carried gear—rucksack, commo, weapon, grenades, and so on—that weighed in at about 110 pounds. These men were the fittest of the fit, and they were ready to do whatever it took to kill the al Qaeda, Taliban, and Chechens wherever they might meet.

The Chechens were the least known of our targets in Afghanistan. The Taliban and al Qaeda were obvious enemies, but the Chechens were a formidable foe. Ousted from their home base in Russia, they were in Afghanistan to assist in the battle against the United States. They were excellent soldiers, good in combat, and well disciplined. Most of our encounters with the Chets were in Lowgar and Paktia provinces. They had no home to return to, so they generally accepted death on the battlefield as their goal. In keeping with that idea, they would fight to the death, and we were happy to assist them in this cause. After kicking their asses, we were left with a healthy respect for these fighting men.

No discussion of Afghanistan can continue without mentioning the weather. As we gained

altitude on our drive on the hardtop road from
Kabul to Lowgar Province—altitude sixty-five
hundred to nine thousand feet at a latitude of 33°
North, 71° east—it became as cold as I could
possibly imagine.

In fact, an old boy from Texas might have
considered it even colder than that.

Cold beyond the possible.

Or, if you prefer, impossibly cold.

I don't wish to bring attention to the following
fact, but I believe I must: I absolutely cannot
stand cold weather. It permeates my bones and
brings out the worst in me. I get irritable and
miserable, and I can't imagine how anyone can
stand it.

But, as I reminded myself repeatedly, I asked
for this war and I was determined to continue
the march. I was childlike in my enthusiasm for
these new weapons, and I wouldn't have traded
the experience for anything. Working with these
bright, young, and strong Americans was such a
great pleasure to me. Above all else, I have always been a man who *functions,* and these lads
functioned.

That said, I was a mess. We dressed like the locals and attempted to pass for them, and in
Afghanistan that meant one thing: a beard. My
beard was a terrible thing, sprinkled with dirt,
snot, and a good amount of food. I had my *chitrali* (Afghan hat) set at a jaunty angle, and at
some point realized it smelled pretty rotten.

When I think of Afghanistan, I think of the

smell—the horrible smell of unclean human beings in subzero temperatures. I am certain every man with us smelled like a cesspool. Personal hygiene went to hell in a handbag, and I was definitely a part of it.

We went more than a couple of weeks without a shower, and everything I owned smelled like what I imagined a rotten asshole would smell like. I know that's a pretty raunchy description, but it's true. We had some cold-assed water, which ran occasionally, but mostly water was scarce. We received an airdrop of water one day, and I watched as a good percentage of the plastic bottles split and sent the water disappearing into the thirsty desert. In that same drop I found some Windex bottles that hadn't split, so I got the bright idea to use the Windex to clean my body.

Squirt, squirt—on with the Windex. And, so I thought—off with the stench.

Bad idea.

I discovered something about Windex: It will take off a couple of layers of skin PDQ. I squirted it only a few times, and every spot of skin that was sprayed with Windex burned like crazy.

After three weeks, my -20°C sleeping bag smelled like the south end of a northbound cow. That goddamned bag was the only place I could be warm. I hated to crawl into it because of the smell, but I damned sure crawled into it, wearing long johns and a wool shirt, snot running freely out of my nose.

My nose ran constantly in Afghanistan. It

dripped into my mustache and then froze in place. I looked like hell, frozen with a green mustache. The only consolation: Everyone else looked the same.

You just learn to live with it, I guess, knowing there is a job to do, knowing there are better times ahead.

The good part? Lice can't live at -10°C.

To sum it up, I was freezing and filthy. My bullet-riddled bones—especially my left knee—ached like mad. A noted junk-food dog, I lost weight as if it was peeling off my body in panels. I was seventy-two years old, fighting a war. All things considered, I wouldn't have traded it for the world.

On a typical day in the Lowgar Province, I got up at 0400 and patrolled our defenses. The SF lads liked this, for they knew I was wary of surprise attack. We had a 24/7 perimeter defense system set up, but sometimes this becomes lax as the weeks pass. After checking the perimeter a few times, I would set up a little kerosene heater and warm my ass up some with that. Then I would roll my bag and prepare my gear.

I set out to do whatever needed to be done. This was my approach to this mission. I knew these younger men were the future of modern warfare, and I was soaking up the information these men took for granted. I checked the geocoordinates and ran the various programs we had on our Falcon View map software. I called the aerial reconnaissance outfit to see if we had any problems or to receive encrypted calls from them to us.

All this took place before daylight. When the sun broke above the mountains we would discuss the plan of the day and work with the anti-Taliban forces at hand. There was one constant, and that was the cold. We stayed cold all fucking day and all fucking night.

Our group hired some Pastuns who turned out to be pretty fine men, with some training. The troops were under the command of General Lodeen, who moved back and forth from Pakistan to Afghanistan, and his son Zia Lodeen, who stayed with us and took charge of the troops.

The Pastuns were beholden to their leader, period. These men were from the southwest of Afghanistan, meaning from Kabul south and west—Pastun country. Many times these men did not mix with the tribes from the northern and northeastern sections of Afghanistan. You see, none of this was run like a republic as we in the United States know it. There were no laws that governed the entire country. There was no leader, now or before, that governed in a manner to which the entire nation adhered. There was tribal law, upheld in different sections of the country as dictated by distance and mountain ranges.

A majority of our Pastun or Dari soldiers could not read or write, nor did they care to learn. These men were fierce in appearance and nonfanatically Islamic in religion. They asked little from their leaders and were armed to the teeth. They took orders only from their number one leader, a common drawback among Islamic armies. Once again, there were no NCOs who, when operating in small groups, could make decisions. All decisions had to come from the top. This hierarchy—no junior leaders, only the boss and the troops—is a fatal organizational mistake, as shown by the crumbling of Saddam Hussein's Iraq.

We were under no illusions regarding the loyalty of these Pastun troops. Each of these men would have cut our hearts out if ordered by their warlord, and they would have kept on cutting until their warlord ordered them to stop. To pay the leader and keep him happy was to keep the army he provided happy. The equation, like the men governed by it, was exceedingly simple.

After four weeks we moved with our indigenous troops south into Paktia Province, about ten kilometers southeast of the city of Gardez. The idea of the war was to control the entire country with the unlikely trinity of Special Forces, the agency, and the warlords. This area of the country was under the control of the warlords, who were kept happy by the agency and SF. This is a different strategy from that used in the Iraq war, where the indigenous Kurds were used only in the north and northeast of that country.

The city of Gardez (a hundred-thousand-plus inhabitants) was loaded with Taliban who had simply changed into civilian attire. The Taliban, as we learned, seldom wore uniforms. This meant everyone was in civilian attire, including the ODA and us. Good guys, bad guys—everybody dressed the same. I believe the Taliban and al Qaeda in the area smelled better than we did, though, for there was a river running on the west side of the city.

We swept into Gardez like a cloud of dust. The city was spared a battle as we moved in and started taking down the big shots. The Taliban didn't even have a chance to shave off their long beards, and the presence of those beards marked them as obvious targets. We had done some advance work—paying off the leaders of the city and bombing out the Taliban hard-liners with smart bombs.

The civilians in Gardez expected the worst, and they were amazed when we came through like a Memorial Day parade. We took the city by storm, our convoy of twenty-five vehicles seeking out the bad guys and rousting them out of their houses. Our cause was aided by the constant circling presence of B-52, B-1B, and B-2 bombers carrying their smart bombs. These aircraft, with their distinctive contrails, were easily seen in the Afghan skies. The Taliban badasses might have been primitive, but they knew they'd eat some of those bombs unless they complied with our wishes.

The coordination was a momentous achievement. The Predator unmanned aerial vehicle (UAV) provided continuous information. We had C-130 Specter gunships available. A look into

the sky revealed never-ending circling of contrails, like so many vultures ready to pounce on
call. Our TACP USAF lads were on their communication devices nonstop, chatting with both
the airborne CP and the combat Talon C-130
gunships.

Our leader paid off an Arab caretaker of an
old mud fort outside Gardez, and in we moved.
The place was large enough for us to store our
important gear, including team vehicles, inside.
The Pastuns set up a perimeter outside the building.

The desert prevailed in this landscape. Gardez
was mostly treeless, with a few scrub trees in our
vicinity. The banks of the river on the west side
were the only areas where trees were prevalent.
This fort was a significant upgrade from the
school building in Lowgar. There was a well inside the mud walls, and after weeks without
ready access to water, this was a godsend. It was
a deep-water well, but I didn't mind cranking
that old handle for twenty minutes to bring some
water up the shaft from about a hundred feet below the surface.

The media attempted to make contact with
our unit many times. We ordered the warlord's
son and Pastun chief, Zia Lodeen, to turn back
the media three kilometers from our location.
This had the desired effect. If a mean-looking,
badassed, smelly Pastun sticks an AK-47 in your
nose, weapon clicked off safe, with his finger on
the trigger, and says in his language, "Hit the
road, Jack" . . . well, you hit the fucking road, is

what you do. Pastun is a difficult language, but those who dealt with Lodeen had no trouble at all understanding the meaning of his words.

Afghanistan was like no other war. Our team set up target locations and received information from high-flying aircraft and other classified means. We knew the direction of the enemy— 360°, in fact, making strikes against the enemy very easy. Our intelligence informed us that in the pass between Gardez and Khost, the last large city in southeastern Afghanistan, there were enough al Qaeda and Taliban to keep us occupied for weeks. Another team inserted into the Khost area lost one U.S. KIA and one U.S. WIA the first night on the ground.

This was not a step-by-step, stop-and-fire type of fight. We moved quickly, with a heck of a lot of air cover, to the next setup, to gather information and ferret out the bad guys. This is the same technique that was used in Iraq, especially in the northern and eastern portions of that theater, where tanks and embedded journalists were not rolling in prime time toward Baghdad.

During a visit into Gardez I accidentally came upon two uniformed Afghans—huge men, with the enormous beards so common among the Taliban. They assured me, through our overworked language man, they were not Taliban. Oh, no. Anything but Taliban.

"We *love* Americans," they both said with wide smiles.

Their eyes weren't smiling, though. Through

those false smiles I could see the hate in their eyes. Of course they were Taliban. Upon further questioning they all but admitted as much, and we were able to garner some beneficial information from these men.

That chance encounter was typical of the fight to rid Afghanistan of the Taliban and al Qaeda. It was not always easy to identify good guys from bad guys. For this reason and others, we did not loiter in Gardez. It had been a Taliban haven, and even though we rolled through its streets on arrival in the area, it was not a place we wanted to roam. We found it more efficient to remain in the fort area, where plans for attacks were initiated.

I spent a lot of time in Afghanistan shaking my head at the phenomenal capabilities of the new weapons systems. I had a hard time adjusting to the hands-off style of warfare, but the Special Forces boys sure enjoyed it. The way it works now, they can drink their coffee while they're destroying somebody.

Without getting too technical, I'd like to outline the workings of the laser-guided weapons systems as we used them in Afghanistan. Our team set up with either a SOFLAM laser-making device or a set of laser Viper binoculars in a position that intelligence indicated the presence of Taliban. At times we were above a spot in the road we knew Taliban would soon cross, or above a house known to be occupied by the enemy.

The next step is to alert the airborne CP that

a target was located and deconfliction—the knowledge that target personnel are not friendly, nor is it civilian—has been established. The laser team paints the target with a SOFLAM infrared laser shot, which bounces off the target and causes "laser splatter," creating a halo effect invisible to the naked eye. This halo is what the smart bomb will seek. This is considered a communications handoff from our team to the bomber, possibly a B1-B carrying laser-guided munitions.

The USAF personnel are masters of this procedure. The bomber talks to the team directly, relating the geocoordinates via shared satellite SATCOM capability. The laser remains trained on the target and the bomb is dropped. These remarkable bombs, their power supplied by gravity, guide themselves, from a height of approximately ten thousand meters, directly into the window of a building or the bed of a pickup truck.

Bam! No more target.

Even smarter are the bombs that guide themselves to an exact position on the ground by a Global Positioning System (GPS). The Taliban could not reason with the Special Operations techniques. Their troops had no answer for it. They couldn't see us, but we were there. Their every movement was monitored; every telephone call was intercepted; every night movement was viewed via night vision devices. There was nowhere to hide. Either the Taliban was viewed from a silent UAV, such as the Predator, or by equally silent Special Forces' team members, viewing targets through night vision devices, or receiving target positions from the aircraft above. What a wonderful combination. This is exactly as I had envisioned Special Operations warfare along the Ho Chi Minh Trail so very many years ago. One problem, though: The technology didn't exist for us then. But lo and behold, our research people had made this happen. The only place the Taliban could remain—and remain undetected—was in their homes, buildings, or caves around the area. And in that case, they still needed to make sure they weren't initiating telephone calls.

The Taliban warning should have been, "You call, you fall," for their airwaves belonged to us.

At roughly 0100 one morning in Gardez—keeping track of dates was not a priority—I received a call from the Washington, D.C., area, from someone watching through a Predator's camera and guiding the UAV to its targets.

The voice related a specific set of geocoordinates and asked, "Do you have any people located at these coordinates?"

I plotted the spot and knew it was an enemy location. I contacted the warlord and others and learned there were no friendlies anywhere near those coordinates.

The Predator relayed a photo of the location back to its station in the United States and the Secure Telephone Unit (STU) III person told me, "We have two sentries walking around a build-

ing at [those same geos]." We established deconfliction and permission was granted to clear the
target for bombing. The Predator relayed a message to a B-1B bomber and the building was
blown to shreds. Later, a small Bomb Damage
Assessment Team (BDA-T) passed through the
area to determine the damage and capture remaining personnel, weapons, and matériel.
These sorts of strikes disorganized the Taliban to
the point where it caused them to abandon the
battlefield completely.

On January 5, sources informed us that a Taliban big-shot meeting was to take place at a certain time in a certain building in Gardez. The
informant told us every big wheel remaining in
the Taliban would attend this meeting. The
Predator, flying soundlessly over this area,
viewed the meeting site. When the appointed
time arrived, several dozen men entered the
building. Then, for several minutes, the Predator
viewed no further entry into the building. This
signified to us that the meeting had commenced.
The Predator received deconfliction and fired its
own Hellfire missile into the building, setting it
aflame.

This was like throwing a rock at a beehive.
There were Taliban running from that building
in a 360° pattern. About twenty of them, armed,
massed about two hundred yards from the
building—bad idea. Another Hellfire missile
landed in the group, leaving about seven or eight
still alive.

The surviving Taliban folks commandeered a

tractor and were hanging off every side of the damned thing, like a backwater version of Dillinger escaping in his touring car.

In the meantime, the Predator relayed a call to a B1-B bomber armed with laser-guided bombs, to this specific area of Gardez. The B1-B entered the area, and the Predator laser-painted the slow-moving tractor. The B1-B released its laser bomb, which hit the tractor dead center and killed all who were riding—or hanging—in it or on it.

The BDA of the building revealed several KIA Taliban and a large cache of weapons. At the risk of repeating myself, our weapons are mind-boggling in their accuracy. I can recall calling fifteen hard HE bombs on a bridge target on the Ho Chi Minh Trail and still not destroying the target. These modern bombs, with their one-meter error factor, would take out that bridge with one bomb, no problem at all.

This technology is so advanced that we could stick a bomb into the opening of a cave, where the thermobaric bomb would detonate only after passing through the adit of the cave. This detonation would cause all the oxygen in the cave complex to be thoroughly depleted. We could send a GPS-guided bomb—through satellite readouts—directly into the front door of an enemy household. We could stick one into the window of a truck. Not only that, but we could choose the window. Front passenger . . . boom! Driver's side . . . boom!

As I watched the Taliban meeting break up

with a well-placed Hellfire missile, then the group of Taliban cut in half by another, then the tractor disintegrate with the B1-B strike, I shook my head and thought about the years we spent in Vietnam, not knowing where the hell we were most of the time, going at the NVA bastards in hand-to-hand combat more often than I'd like to remember. We used mirrors to summon dumb bombs to the vicinity of the targets, hoping the damn things wouldn't land on us. So many of our fine young SOG SF men were KIA after facing off with the enemy in close combat, and I thought about how close we were to figuring this whole thing out. We had the concepts of small-team infiltration down pat, but we were short on the technology. *Damn,* I thought, *our old SOG recon patrols were ahead of their time. We had the right idea, just not the right tools.*

As we roamed the area surrounding Gardez we became increasingly curious about the prevalence of little yellow-and-orange taxis grinding through the hills back and forth from the border with Pakistan. In time we realized these ten-year-old Russian-made Ladas could only be hauling Taliban and al Qaeda out of the battle area. It seemed these guys had established a cave-to-border service.

We initiated a halt to all traffic, and many of the Taliban and al Qaeda were apprehended in this fashion, making us wonder how many had slipped through before we instituted our checkpoint. Word traveled through the Taliban grapevine, and the taxis quickly sought routes inaccessible to even a tank. These taxi drivers, like taxi drivers everywhere in the world, were an enterprising lot.

The impossibility of covering each and every route along these imposing ten-thousand-foot to twelve-thousand-foot mountain passes no doubt resulted in our missing important Taliban. However, Afghanistan is like west Texas; to cover every trail, route, road, and path was beyond human capability. Some of the Taliban were able to break free of our web and escape into Pakistan.

As an aside, it is my opinion that Usama bin Laden was killed in action on February 4, 2002. I believe he was reduced to very small DNA, compliments of a very smart bomb guided by a Predator. I know this is open to debate, since his DNA was not found on the bomb site in the Tora Bora area. My opinion is based partly on knowledge and partly on the feeling that this man could not stay away from the camera. Even on his deathbed, he would be making live TV appeals for funds to counter the Americans.

In those mountain passes on the Pakistan-Afghan border, the Pakistani troops manning the customs checkpoints stayed inside their huts most of the time, huddled near their kerosene stoves trying to keep warm. Meanwhile, the taxis were stopping in the middle of the desert and releasing all their al Qaeda passengers to cross the border undetected.

One day I stopped a taxi with no passengers.

"Who are your normal passengers?" I asked the driver.

He made some motions with his hands and told me some bullshit or other. These Pakistani hustlers would lie through their teeth to protect their scam. I decided to take a different approach.

Pointing to the nearby mountains, I asked, "How much for a ride from those caves to the border?"

He fussed around but finally told me: $85 U.S. or a washtub full of Pakistani rupees. I pressed him, and he told me he often jammed six of these stinking folks in his cab as he crawled along the dirt roads in the battle area to the larger paved roads, then on toward Pakistan.

During my two months in Afghanistan, I managed to avoid pneumonia and gunshot wounds. Despite my history of attracting gunshots, avoiding pneumonia was my greatest feat. By the end of those two months, I had shed thirty-five pounds from my old body. I was cold, filthy, and stinky, but I was one of roughly 150 men who can say they conducted combat on the ground in Afghanistan during the initial—and pivotal—phase of Operation Enduring Freedom.

I have to admit, though, this standoff type of battle did not appeal to me. We won quickly and lost few men in battle, so its success cannot be argued. I thought back to thirty-six years earlier, in the rice paddies of Bong Son, fighting hand-to-hand and dodging bullets—most of them, anyway—in an effort to survive. I can't say I missed that type of battle, but this new one felt anticlimactic.

During the middle of January 2002, I completed my two-month tour in Afghanistan and said good-bye to the fine men of ODA 594. They treated me with great deference and respect, and I left them knowing the proud tradition of SF was in good hands.

I left the battlefield and made my way back to the United States, through Uzbekistan, Azerbaijan, and Germany.

The last battle. The last time with field troops.

Life continues. The young and strong take the place of the old and weak.

I understand.

But damned if it didn't make me terribly sad.

I thought back to the desperation of Bong Son, the helplessness of Oscar-8, the pride of Ba Kev. So many battles, so many men lost and saved, so many life-or-death decisions forced on me. I thought back to the lonely hours spent watching bin Laden and tracking Carlos and dealing with so many others I can't mention.

I had made the edge my home for more than fifty years, and like an athlete who lives for the rush of the crowd's approval, I knew it would be difficult to live without the adrenaline of combat or espionage.

The memories swept through me.

It was somebody else's hunt now.

AFTERWORD

I wrote this book to shed light on my participation in some wonderful actions over a fifty-year period. Many of these actions have remained unreported and undercover until now. I believe these great operations, and the men who conducted them, deserve to be recognized.

Hero is not a term I use loosely. In fact, heroism has never entered my mind, not on the battlefield or in an observation post. A good combat man, as I see it, is one who has control of his wits, regardless of the hellish action surrounding him. Men are not heroes by birth, but circumstances can thrust a man into a position to respond in a heroic fashion. Through the years I have seen men respond as heroes, to think, act, and react quickly enough in combat to perform heroic acts.

To me, the essence of heroism is defined in Paragraph II of the U.S. Army Field Operations

Order: "TO COMPLETE THE MISSION, regardless of the enemy counteraction, and keep his men alive."

This book is for those men who have honored that code.

Many of their acts are documented on the preceding pages, but there was not room for all of them. This does not make their heroism any less significant. I have witnessed heroism from agency personnel, on the streets of Khartoum and in countless other countries. But in my heart I am a combat man, a Special Forces man, and I have a particular fondness for the men who performed with extreme valor on the ground and in the air during my seven and a half years in Vietnam.

I've witnessed U.S. pilots, of fixed-wing aircraft and helicopters, act with bravery and valor beyond measure. Pilots who would fly through a wall of green-tracer steel to save a Special Forces soldier or Montagnard who otherwise would have died.

I have witnessed our SOG mercenary pilots, from the 219th VNAF, who constantly astounded me with their lack of regard for their own safety—in a word, bravery. These men flew their helicopters through the jungle along the Ho Chi Minh Trail, chopping branches and entire trees as they strived to save a recon team on the ground.

I once witnessed Mustachio—our finest Vietnamese chopper pilot—take a gunshot wound to the neck, then place his knuckles into that gaping wound to stop the bleeding as he flew his H-34 out of the battle area. Mustachio's uncommon valor and total disregard for his own well-being saved the life of SOG Sergeant First Class Harry Brown.

I watched as Staff Sergeant Lester Pace lowered himself into the battle area of O-8 to pick up Sergeant First Class Charles Wilklow, who despite being gravely wounded managed to evade the NVA for several days.

Did Lester Pace argue with me about completing this action? Not on your life, for he was a Special Forces man and a SOG man. He did as he was told, and he completed the mission.

I have seen things I'd rather forget. One day in South Vietnam I stood next to Sergeant First Class Bruce Luttrell as we discussed a plan of action. Without warning, he was hit by shrapnel that pierced his skull and entered his brain. I dragged his body out of the line of fire, but this fine man died of his wounds. War is maddeningly random.

I know an H-34 pilot named Captain Nguyen An, who somehow lowered his H-34 helicopter over a rushing river and lifted a downed HU-1D helicopter with three men aboard using only the wheels of his own bird. An was under relentless enemy fire, but this downed chopper was about to be washed away in the river. Disregarding the onslaught of NVA might, An completed this phenomenal feat of dexterity and bravery, saving the lives of three U.S. servicemen in the process. Captain An later lost both hands when he was almost burned to death in another rescue attempt on the Ho Chi Minh Trail.

I'll never forget the unbelievable story of Staff Sergeant Sammy Hernandez, a man I rescued personally. Hernandez was dangling beneath a helicopter by a climbing rope when he took on enemy fire. The climbing rope snapped and Sammy was dropped onto the battle area, alive but stunned. All seven other Americans were killed instantly, but Sammy hung on and kept his cool, escaping and evading the enemy for a lengthy period. I eventually rescued him after he signaled me with his trusty mirror. Sammy Hernandez, a true Special Forces man, continued combat actions for several years and was a member of the first combat HALO insertion team.

So many missions, so much bravery. One day I told Sergeant First Class Jimmy Scurry, a Special Forces medic, to grab his gear.

"We're going on a Silver Star rescue mission," I told him.

Scurry immediately started to get himself ready and said, "Damn, Billy, let's make it the Distinguished Service Cross."

Jimmy Scurry kept his wits in combat. Jimmy Scurry displayed the kinds of qualities we associate with heroes.

After Mustachio was shot through the neck and landed his bird safely in Khe Sanh, I watched Scurry carry the great pilot into a U.S. emergency military hospital tent. Inside, a military medical man told Scurry, "We don't treat Vietnamese in this tent."

Jimmy Scurry grabbed the medical man by his lapels and said, "You're going to treat this man, or I'll whip your ass right out of this tent."

Jimmy Scurry was very large and very mean. Mustachio was treated immediately.

Mustachio lived to fly again. He was shot down and taken prisoner. He never returned from captivity. Among the men whose lives he saved, he will remain a legend, never to be forgotten.

In 1963, I watched an SF sergeant first class perform a feat of heroism while crossing a flooded stream under enemy fire. One of his Montagnards, a nonswimmer, was deep under water and quickly drifting downstream. This SF man disregarded the fire, tossed aside his weapon, and rescued this Montagnard, pulling him to safety and breathing life back into his lungs.

I asked a lot of my men, and they responded accordingly. I remember telling Manny Bustemante to drop into the battle area from a CH-53 hoist, alone, to search for a missing SF recon man named Flora. We didn't know it, but Flora had been captured by the enemy. Did Bustemante argue or hesitate? Not on your goddamned life. He went into the battle area, alone, and attempted to pick up Staff Sergeant Flora.

I witnessed an F-4C Phantom get blasted from the skies over War Zone D by an NVA antiaircraft weapon. The pilot ejected, and Master Sergeant Henry Bailey and I jumped into a Support HU-1D rescue bird, went airborne, and headed toward the parachuting pilot. This pilot's parachute had caught on a tree limb, at least 150 feet above the jungle.

As the NVA scrambled to get to the pilot, the HU-1D moved to the pilot's side as he hung from the tree. We managed to snap-link the pilot onto the rescue ladder, chop his suspension lines, and fly him away from certain death. The most amazing thing? That pilot's feet never touched the ground until he was in a friendly landing zone.

Our best POW snatcher was a recon man named Sergeant Jenkins. He once ordered an H-34 to land among buildings, in broad daylight, along the Ho Chi Minh Trail. Jenkins then hopped out, ran to a hooch, kicked the door in, grabbed an NVA, dragged him and his AK-47 out, and tossed him onto the helicopter. Away they flew with this POW—not a shot was fired.

Heroes are made of men who stay focused in battle. Heroes are men who have a plan, and who understand the plan is ever-changing and dependent on enemy action. Heroes are those who can adjust to a new plan and then execute that plan even when the enemy interferes with his intentions.

I was in and around U.S. Army Special Forces

from 1954 to 2001. We had to fight to remain in existence in 1956, then again in 1972, when the generals at the highest levels of the U.S. Army wished Special Forces to be abolished.

Special Forces teams have been deployed to more than 130 countries. They have taught, led, and cared for indigenous. They have completed wonderful operations and continue to do so today. Their exploits have turned the old Special Forces–hating generals into retirees.

Now Special Operations Command (SO-COM) has a four-star general officer and its own budget separate from the regular military budget. It has its own air and naval support. It has taken the Force Marine units into its fold and is ready and willing to do Special Operations versus anyone the administration and DOD wish to deal with.

I have never heard a Special Forces soldier say, when assigned a very difficult mission, "I am not going to do that."

Instead I have always heard the SF soldier say, "Let's get it on. Let's go."

An SF man's motto is "Lead, follow, or get the hell out of the way."

There is a wonderful stone near the Tomb of the Unknown Soldier in Arlington Cemetery. It is dedicated to the Special Operations Forces. This marble stone, flush to the ground, sits under a large oak tree.

Etched on this beautiful gray marble stone are the words from Isaiah 6:13, which read: "And the Lord said, who will go, who will fight for me," and the young Special Forces man stepped away from his family as they reached out, saying, "I will go, send me."

ACKNOWLEDGMENTS

So many people helped both in the creation of this book and the life it describes. What follows is undoubtedly an incomplete list of those who deserve my acknowledgment:

My niece Suzanne Sanders and her mother (and my sister) Nancy Sanders, who have always been so helpful and positive; Billie Alexander, who listened to me and let me win; Martha Raye; Dr./Colonel Arthur Metz, who saved my right foot and leg when it didn't seem possible; my Special Forces friends, who were my mentors and taught me so much.

I owe a debt of gratitude to Cofer Black, a no-nonsense, straight-shooting senior intelligence supervisor. Along those lines, I would also like to acknowledge all the fine, fine people with whom I have worked from the CIA. You are part of an excellent outfit, and never let the bastards get you down. I wish I could mention your names, but—

as you lads and lasses know better than any-one—I surely cannot.

Thanks to Tim Keown, the ghost for this book; my best to your lady, Miriam, and your four fine boys. And I offer a crisp salute to Doug McMillan, who was responsible for bringing us together.

Our fine agent, Frank Weimann, took care of business while our editors, Mauro DiPreta and Joelle Yudin, believed in the project from the start.

Special thanks to Nancy McCarthy and Eileen Knight for scurrying about and helping with photos, while I "vacationed" in Baghdad. You really came through for me.

And finally, to the Special Forces and the fine young men who have been killed in action. You have been my life for so long, I will work at our job until I join you.

TRUE AND INTIMATE PORTRAITS OF AMERICA'S WARRIORS ON THE FRONT LINES OF BATTLE

One Perfect Op
Navy SEAL Special Warfare Teams
by Command Master Chief Dennis Chalker, USN (Ret.)
with Kevin Dockery
0-380-80920-6/$6.99 US/$9.99 Can

Delta Force
The Army's Elite Conterterrorist Unit
by Col. Charlie A. Beckwith (Ret.)
and Donald Knox
0-380-80939-7/$7.99 US/$10.99 Can

Big Red
The Three-Month Voyage of a Trident Nuclear Submarine
by Douglas C. Waller
0-380-82078-1/$7.99 US/$10.99 Can

Good to Go
**The Life and Times of a Decorated Member of
the U.S. Navy's SEAL Team Two**
0-380-72966-0/$7.50 US/$9.99 Can

Hunting the Jackal
**A Special Forces and CIA Soldier's
Fifty Years on the Frontlines of the War Against Terrorism**
by Billy Waugh with Tim Keown
0-06-056410-5/$7.99 US/$10.99 Can